Shahab Karrari

Integration of Flywheel Energy Storage Systems in Low Voltage Distribution Grids

Integration of Flywheel Energy Storage Systems in Low Voltage Distribution Grids

by

Shahab Karrari

Scientific Publishing

Karlsruher Institut für Technologie
Institut für Technische Physik

Integration of Flywheel Energy Storage Systems
in Low Voltage Distribution Grids

Zur Erlangung des akademischen Grades eines Doktor-Ingenieurs
von der KIT-Fakultät für Elektrotechnik und Informationstechnik des
Karlsruher Instituts für Technologie (KIT) genehmigte Dissertation

von Shahab Karrari, M.Sc.

Tag der mündlichen Prüfung: 24. Februar 2021
Hauptreferent: Prof. Dr.-Ing. Mathias Noe
Korreferent: Prof. Dr.-Ing. Marc Hiller

Bild Umschlagvorderseite: KIT, Institut für Technische Physik (ITEP)
Fotografie: KIT, Amadeus Bramsiepe

Impressum

KIT Scientific Publishing

Karlsruher Institut für Technologie (KIT)
KIT Scientific Publishing
Straße am Forum 2
D-76131 Karlsruhe

KIT Scientific Publishing is a registered trademark
of Karlsruhe Institute of Technology.
Reprint using the book cover is not allowed.

www.ksp.kit.edu

Print on Demand 2022 – Gedruckt auf FSC-zertifiziertem Papier

ISBN 978-3-7315-1140-3
DOI 10.5445/KSP/1000139457

Acknowledgements

First and foremost, I would like to express my sincere appreciations to my supervisor Prof. Dr.-Ing. Mathias Noe, for his continuous and consistent support throughout my entire PhD, and for his trust in me to pursue my own research ideas.

Next, I would like to thank the DFG research training group "Energy Status Data" for providing the opportunity to carry out my doctoral studies at KIT, and for establishing a platform for interdisciplinary discussions with other researchers. I am very grateful to all the faculty and graduate members of the research training group for the fruitful discussions and constructive feedback during our meetings. In particular, I would like to thank Nicole Ludwig and Michael Vollmer, who have made parts of this work possible through effective collaborations.

I would like to thank Dr. Giovanni De Carne for his support during the last year in reviewing manuscripts and the dissertation, and providing productive feedback for my work. I also thank Dr. Jörn Geisbüsch for proofreading the manuscripts.

The experiments conducted at KIT's EnergyLab 2.0 would have not been possible without the generous help of Mr. Christian Lange and Mr. Frank Gröner. Christian somehow managed to always find a simple and yet effective solution to any hardware issues faced during experiments.

The grid measurements were collected with the cooperation and kind assistance of Mr. Christian Seiler and Mr. Peter Friedmann.

I would like to thank all my colleagues and officemates at ITEP, in particular, Roland Gyuráki, Ruslan Popov, Fabian Schreiner, Nicolò Riva, Philip Kreideweis, Cyra Neugebauer, Tara Benkel, Dustin Kottonau, and Wescley de Sousa. Thank you for being more than just colleagues and hearing me out in times of trouble.

Lastly, I would like to thank my family for their remote but unwavering support during this rather challenging part of my life.

Karlsruhe, December 14, 2020

Shahab Karrari

Kurzfassung

Mit dem Ziel, den Stromsektor zu dekarbonisieren und dem Klimawandel zu begegnen, steigt der Anteil erneuerbarer Energieressourcen in den Energiesystemen rund um den Globus kontinuierlich an. Aufgrund des intermittierenden Charakters dieser Ressourcen kann die Aufrechterhaltung des momentanen Gleichgewichts zwischen Erzeugung und Verbrauch und damit der Netzfrequenz ohne angemessene Maßnahmen jedoch eine Herausforderung darstellen. Da erneuerbare Energiequellen mit Umrichterschnittstellen dem System selbst keine Trägheit verleihen, nimmt gleichzeitig die kumulative Systemträgheit ab, was zu schnelleren Änderungen der Netzfrequenz und Bedenken hinsichtlich der Netzstabilität führt.

Ein Schwungrad-Energiespeichersystem (Flywheel Energy Storage System, FESS) kann schnell große Leistungsmengen einspeisen oder aufnehmen, um das Netz nach einer abrupten Änderung der Erzeugung oder des Verbrauchs zu unterstützen. Neben der schnellen Reaktionszeit hat ein FESS den Vorteil einer hohen Leistungsdichte und einer großen Anzahl von Lade- und Entladezyklen ohne Kapazitätsverlust während seiner gesamten Lebensdauer. Diese Eigenschaften machen das FESS zu einem gut geeigneten Kandidaten für die Frequenzstabilisierung des Netzes oder die Glättung kurzfristiger Leistungsschwankungen auf lokaler Ebene.

In dieser Dissertation wird die Netzintegration eines Hochgeschwindigkeits-FESS auf der Niederspannungsebene aus mehreren Perspektiven untersucht. Zunächst wird das Problem der Platzierung und Dimensionierung eines FESS in Niederspannungsverteilnetzen für Leistungsglättungsanwendungen behandelt. Um den am besten geeigneten Standort für ein FESS zu finden, wird eine datengetriebene Methode zur Abschätzung der relativen Spannungsempfindlichkeit vorgestellt, die auf dem Konzept der Transinformation basiert. Der Hauptvorteil der vorgeschlagenen Methode besteht darin, dass sie kein Netzmodell erfordert und nur Messwerte an den interessierenden Punkten verwendet. Messergebnisse aus einem realen Netz in Süddeutschland zeigen, dass mit dem vorgeschlagenen Ansatz die Netzanschlusspunkte mit einer höheren Spannungsempfindlichkeit gegenüber Wirkleistungsänderungen, welche am meisten von einem durch FESS ermöglichten, glatteren Leistungsprofil profitieren können, erfolgreich zugeordnet werden können. Darüber hinaus wird eine neue Methode zur Dimensionierung von Energiespeichersystemen unter Verwendung von Messdaten eingeführt. Der vorgeschlagene Ansatz erkennt wiederkehrende Verbrauchsmuster in aufgezeichneten Leistungsprofilen mit Hilfe des "Motif Discovery"-Algorithmus, die dann zur Dimensionierung verschiedener Speichertechnologien, einschließlich eines FESS, verwendet werden. Anhand von gesammelten Messdaten aus mehreren Niederspannungsnetzen in Deutschland wird gezeigt, dass die Speichersysteme mit den aus den detektierten Mustern abgeleiteten Charakteristika während der gesamten Messperiode effektiv für ihre Anwendungen genutzt werden können.

Als nächstes wurde ein dynamisches Modell eines Hochgeschwindigkeits-FESS entwickelt und mit experimentellen Ergebnissen in mehreren Szenarien, unter Berücksichtigung der Verluste und des Hilfsenergiebedarfs des Systems, validiert. In den untersuchten Szenarien wurde eine maximale Differenz von nur 0,8 % zwischen dem Ladezustand des Modells und dem realen FESS beobachtet, was die Genauigkeit des entwickelten Modells beschreibt.

Nach Festlegung des erforderlichen Aufbaus wurde die Leistungsfähigkeit eines 60 kW Hochgeschwindigkeits-FESS während mehrerer Frequenzabweichungsszenarien mit Hilfe von Power Hardware-in-the-Loop-Tests beurteilt. Die Ergebnisse der PHIL-Tests zeigen, dass das Hochgeschwindigkeits-FESS sehr schnell nach einer plötzlichen Frequenzabweichung reagiert und in knapp 60 ms die erforderliche Leistung erreicht, wobei die neuesten Anforderungen der Anwendungsregeln für die Frequenzunterstützung auf der Niederspannungsebene erfüllt werden.

Um schließlich die Vorteile des schnellen Verhaltens des FESS für Energiesysteme mit geringer Trägheit zu demonstrieren, wurde ein neuartiger adaptiver Trägheits-Emulationsregler für das Hochgeschwindigkeits-FESS eingeführt und seine Leistung in einem Microgrid mit geringer Trägheit durch Simulationen und Experimente validiert. Die Simulationsergebnisse zeigen, dass die Verwendung des FESS mit dem vorgeschlagenen Trägheits-Emulationsregler die maximale Änderungsrate der Frequenz um 28 % und die maximale Frequenzabweichung um 44 % während der Inselbildung des untersuchten Microgrid reduzieren kann und mehrere zuvor vorgestellte adaptive Regelungskonzepte übertrifft. Der vorgeschlagene Regler wurde auch auf einem realen 60 kW FESS mit dem Konzept des Rapid Control Prototyping implementiert, und die Leistungsfähigkeit des FESS mit dem neuen Regelungsentwurf wurde mit Hilfe von PHIL-Tests des FESS validiert. Die PHIL-Ergebnisse, die die allererste experimentelle Validierung der Trägheitsemulation mit einem FESS darstellen, bestätigen die Simulationsergebnisse und zeigen die Vorteile des vorgeschlagenen Reglers.

Abstract

With the aim of decarbonizing the electricity sector and addressing climate change, the share of renewable energy sources in power systems around the globe is consistently increasing. However, due to the intermittent nature of these sources, maintaining the instantaneous balance between the generation and the demand, and therefore, the grid frequency, can be challenging without adequate measures. In addition, since converter-interfaced renewables do not inherently provide inertia to the system, the cumulative system inertia is simultaneously decreasing, resulting in faster changes in the grid frequency and concerns over the grid stability.

A Flywheel Energy Storage System (FESS) can rapidly inject or absorb high amounts of power in order support the grid, following an abrupt change in the generation or in the demand. In addition to the quick response time, a FESS has the advantage of a high power density and a large number of charging and discharging cycles, with no capacity loss throughout its lifetime. These characteristics make the FESS a well-suited candidate for providing frequency support to the grid or smoothing out short-term power variations at a local level.

The work presented in this thesis studies the grid integration of a high-speed FESS in low voltage distribution grids from several perspectives. First, the problem of allocation and sizing of a FESS in low voltage distribution grids for power smoothing applications is addressed. In order to find the most suitable location for a FESS, a data-driven method for estimating the relative voltage sensitivity is introduced based on the concept of mutual information. The main advantage of the proposed method is that it does not require a grid model, and uses only the measurement values at points of interest. Measurement results from several real distribution grids in southern Germany show that the proposed approach can successfully allocated the grid connection points with a higher voltage sensitivity to active power changes, which can benefit the most from a smoother power profile, using a FESS. Moreover, a new method for sizing energy storage systems using historical measurement data is introduced. The proposed approach detects reoccurring consumption patterns in the measured power profiles using the motif discovery algorithm, which are then used for sizing different energy storage technologies, including a FESS. Using collected measurement data from several low voltage distribution grids in southern Germany, it is demonstrated that the storage systems with the characteristics derived from the detected patterns only can be effectively used for their intended applications during the whole measurement period.

Next, a dynamic model of a high-speed FESS has been developed, which incorporates the losses and auxiliary power requirements of the system. The model is validated with experimental results from a real FESS in several scenarios. A maximum difference of only 0.8 % has been observed between the state of charge of the real FESS and its model during the investigated scenarios, reflecting the accuracy of the developed model.

After establishing the required setup, the performance of a 60 kW high-speed FESS during several frequency deviation scenarios has been evaluated by means of Power Hardware-in-the-Loop (PHIL) testing. The PHIL testing results show that the high-speed FESS responds very quickly and reaches the required power in just under 60 ms following a frequency deviation, while complying with the latest grid code requirements for frequency support at the low voltage level in Germany.

Lastly, in order to demonstrate the advantages of the rapid response of the FESS in low-inertia power systems, a novel adaptive inertia emulation controller for the high-speed FESS has been introduced and its performance has been validated in a low-inertia microgrid using simulations and experiments. Simulation results show that the use of the FESS with the proposed inertia emulation controller can reduce the maximum rate of change of frequency by 28 %, and the maximum frequency deviation by 44 % during the islanding of the studied microgrid. In addition, the proposed controller outperforms several previously reported adaptive control designs in terms of limiting the frequency deviations. The proposed controller has also been implemented on a real 60 kW FESS using the concept of rapid control prototyping, and the performance of the FESS with the new control design has been validated by means of PHIL testing. The PHIL results, which are the very first experimental validation of inertia emulation using a FESS, confirm the simulation results, demonstrating the benefits of the proposed controller in low-inertia power systems.

Table of Content

Acknowledgements i

Kurzfassung iii

Abstract v

1 Introduction, Motivation, and Scope of Work 1

2 The Role of Flywheel Energy Storage Systems in Power Systems 7

2.1 Introduction to Flywheel Energy Storage Systems (FESS) 7
2.2 Applications of FESS in Power Systems 10
2.2.1 Frequency Regulation 10
2.2.2 Power Smoothing 12
2.2.3 Voltage Regulation 13
2.2.4 Other Applications in Power Systems 13
2.3 Components of FESS 14
2.3.1 Rotor 15
2.3.2 Electrical Machine 16
2.3.3 Power Converters 17
2.3.4 Bearings 19
2.3.5 Enclosure 21
2.4 Summary 21

3 Real-time Simulation and Power Hardware-in-the-Loop Testing for Grid Integration Studies 23

3.1 Challenges of Real-time Simulations of Low Voltage Distribution Grids 23
3.1.1 Lack of Decoupling Points for Parallel Computing 25
3.1.2 Aggregation of Distributed Energy Resources 26
3.1.3 Presence of Asymmetry and Harmonics 27
3.1.4 A Summary of the Solutions for Real-time Simulation of Low Voltage Distribution Grids 28
3.2 Power Hardware-in-the-Loop (PHIL) Testing of Distributed Energy Resources 29
3.2.1 Test Case Development 31
3.2.2 Stability and Accuracy Evaluation 32
3.2.3 Execution of the PHIL Testing 41
3.2.4 State-of-the-art in PHIL Testing of Energy Storage Systems 41
3.3 Description of the 1-MVA PHIL Infrastructure at KIT 42

3.4 Summary 45

4 Data-driven Allocation and Sizing of Energy Storage Systems 47

4.1 A Data-driven Method for Allocation of Flywheel Energy Storage Systems 48

4.1.1 Introduction to Voltage Sensitivity 49

4.1.2 Limitations of the Classical Method for Calculating the Voltage Sensitivity 53

4.1.3 Introduction to Mutual Information (MI) 54

4.1.4 The Proposed Allocation Method for a FESS 56

4.1.5 Evaluation of the Proposed Allocation Method 61

4.2 A Data-driven Method for Sizing ESS using Standard Patterns 64

4.2.1 The Input Data for the Sizing 68

4.2.2 Introduction to Motif Discovery 70

4.2.3 Standard Patterns for Sizing ESS 73

4.2.4 Deriving Energy Storage Characteristics 76

4.2.5 Allocating Power Profiles 79

4.2.6 Sizing Outcome 82

4.2.7 Evaluation of the Proposed Sizing Method 86

4.3 Summary 88

5 Modelling of High-Speed Flywheel Energy Storage Systems (FESS) for Real-time Simulations 91

5.1 Potential Applications of Real-time Simulation of a FESS 91

5.2 Modeling of a High-speed Flywheel Energy Storage System (FESS) 92

5.2.1 Modeling of a Permanent Magnet Synchronous Machine 94

5.2.2 Modeling of Voltage Source Converters 100

5.2.3 Grid-side Converter Controller 105

5.2.4 Machine-side Converter Controller 109

5.2.5 Modeling of Synchronous Reference Frame Phase-Locked Loop 112

5.2.6 Modeling of System Losses and Auxiliary Power Requirements 112

5.3 Real-time Simulation of the High-speed FESS 115

5.3.1 Frequency Step Response 116

5.3.2 Voltage Sag Response 117

6 PHIL Testing and Model Validation of a High-Speed FESS for Frequency Support 119

6.1 Description of the Hardware under Test 119

6.2 Description of the PHIL Setup 120

6.3 PHIL Testing Scenarios and Model Validation of the FESS 122

6.3.1 Scenario A: Charging at the Rated Power 122

6.3.2 Scenario B: Grid Code Compliance Verification Test for Frequency Support ... 124
6.3.3 Scenario C: Response to the UK's Frequency incident of August 9, 2019 ... 128
6.3.4 Scenario D: Increasing the Frequency Support of a Low-voltage Distribution Grid ... 132
6.4 Summary ... 136

7 Adaptive Inertia Emulation using High-speed Flywheel Energy Storage Systems ... 137
7.1 Introduction ... 137
7.1.1 The Need for Inertia Support from Converter-interfaced Systems ... 138
7.1.2 Flywheels as the Active Power Source for Inertia Emulation ... 141
7.1.3 Inertia Emulation versus Virtual Synchronous Machines ... 143
7.1.4 State-of-the-Art in using a FESS for Inertia Emulation ... 144
7.2 Adaptive Inertia Emulation Control for a High-speed FESS ... 145
7.2.1 Controller Design of Inertia Emulation using High-speed FESS ... 145
7.2.2 Adaptive Inertia and Damping for an Improved Response ... 148
7.2.3 Parameters for the Inertia Emulation Control using a FESS ... 151
7.2.4 Estimation of the Frequency and its Derivative ... 153
7.3 Simulation Results of the Adaptive Inertia Emulation using a high-speed FESS ... 155
7.3.1 Scenario A: Microgrid Islanding ... 157
7.3.2 Scenario B: Drop in the PV Generation ... 161
7.4 PHIL Testing of a High-speed FESS for Inertia Emulation ... 163
7.4.1 Real-time Simulation of the Microgrid ... 163
7.4.2 Implementing the Inertia Emulation Controller on the real FESS ... 163
7.4.3 Scenario A: Microgrid Islanding ... 166
7.4.4 Scenario B: Drop in the PV Generation ... 168
7.4.5 Comparison of the PHIL Testing and the Simulation Results ... 171
7.5 Summary ... 172

8 Conclusions and Outlook ... 173
8.1 Conclusions ... 173
8.2 Outlook ... 175

Bibliography ... 177

Publications ... 205

Appendix A. A Summary of Power Quality Measurements ... 207
A.1 Power Quality Requirements for Low Voltage Grids in Germany ... 207

A.2 Summary of Interesting Recorded Incidents 209

Appendix B. Parameters of the Microgrid Components 213

Table of Figures...215

Table of Tables 223

Abbreviations 227

1 Introduction, Motivation, and Scope of Work

In order to comply with the global efforts to address climate change, more and more renewable energy sources are being integrated into power systems around the world. The global installed capacity of renewables has been steadily increasing over the last decades, reaching approximately 200 GW by the end of 2020 [1]. Renewables are projected to become the largest source of energy for electricity generation worldwide by 2025, surpassing coal, and supplying one-third of the global electricity demand. By 2050, over 80 % of the world's electricity could be supplied by renewables [2]. However, maintaining the stability and reliability of the power systems and the power quality can be challenging with such a high share of renewables.

In Germany, the share of renewables in the electricity supply reached approximately 42 % in 2019, exceeding the target of 35 %, initially planned for 2020 according to the renewable energy act [3]. Solar and wind generation units make up the largest share of renewable generation in Germany, with a current total installed capacity of 45 GW and 59 GW, respectively. By 2034, the net installed capacity of wind and solar generation is expected to reach 173 GW, which is approximately twice the current electricity consumption peak in the country. However, due to the intermittent nature of these resources, the instantaneous mismatch between the generation from renewables and demand is expected to grow as high as 74.3 GW [4]. Therefore, balancing the volatile generation and the demand in each and every moment can be challenging without adequate measures. This balance is reflected in the grid frequency, which increases from the nominal value in case of generation exceeding demand, and decreases in case of insufficient generation. For large deviations of frequency above a certain limit, renewables are required to be curtailed, losing useful energy, while for large under-frequency incidents, large consumers are required to be disconnected by the under-frequency load shedding schemes, leading to significant financial losses.

Furthermore, contrary to conventional power plants, converter-interfaced generation units, such as wind and solar, do not inherently provide inertia to the power system. Therefore, with the increasing share of converter-interfaced renewables, the total system inertia is steadily decreasing in power systems around the globe, including in Europe [5]. This leads to a higher rate of change of frequency, which can potentially lead to an increase in the number of frequency violations and activation of protection systems based on a high rate of change of frequency. This issue is expected to become more critical by the decommissioning of nuclear and coal power plants in countries such as Germany in the near future. With the reduced system inertia in the interconnected European power system, a system split, similar to the one occurred on November 4, 2006, can have serious consequences for the European power system [5], [6]. In smaller power systems with less interconnections, such as the power system of the UK and Ireland, the increase in the rate of change of frequency has already caused great concerns, and precautionary measures are being adopted [7].

One of the major solutions is the use of Energy Storage Systems (ESS), which can help balancing the intermittent power of renewables and the demand in a wide range of time scales, depending on the technology. Storage systems can play a fundamental role in integrating renewable energy resources and regulating the grid frequency. The International Energy Agency (IEA) has estimated that in order to support the electricity sector decarburization plans, at least 310 GW of additional energy storage systems are required only in the four major regions of Europe, United States, China, and India [8]. High investment costs and regulatory and market constraints are considered as the main barriers towards a more widespread use of ESS in power systems [9]. For balancing the generation and the demand, ESS are complimented by other technical solutions, including flexible demand and generation, sector coupling (e.g., heat, gas), and increasing the power transfer capabilities with additional power corridors (e.g., HVDC).

In order to support the grid frequency in power systems with a reduced system inertia, an ESS should be able to respond promptly to frequency deviations caused by a sudden change in the renewable generation or in the demand. Therefore, a fast-reacting ESS with a high power density is required, which can rapidly provide or absorb high amounts of power in order to support the grid during disturbances [7], [10]–[12]. While a handful of storage technologies can provide fast frequency support services to the grid, several studies have shown that Flywheel Energy Storage Systems (FESS) are the most cost-effective solution for high-power short-term ancillary services such as primary frequency support, due to their low operational and maintenance costs and long lifetime [13], [14]. Supercapacitors [11] and lithium-ion batteries [15] are the main competitors of the FESS for this application. However, contrary to the FESS, they can degrade with time and temperature [16], [17], and often particular charging strategies may be required to be adopted to preserve their lifetime [15]. However, a FESS can quickly provide or absorb high amounts of power with no concern over its lifetime or capacity, and support the grid during sudden variations in generation or in demand. With a high number of cycles in its lifetime, a FESS can also be used for smoothing out short-term power fluctuations of renewables or loads at their connection point to the grid, and avoid fast voltage changes caused by active power changes, and the spread of fast power fluctuations to the rest of system.

The focus of this work is to study the integration of a FESS in low voltage distribution grids, where a significant share of distributed and renewable energy resources is connected. Grid integration of new power system components are commonly investigated using either numerical simulations or experimental tests. With the increasing capabilities and wide-spread availability of real-time simulators, Hardware-in-the-Loop (HIL) and Power Hardware-in-the-loop (PHIL) testing are becoming one of the major tools for grid integration studies of new power system components and validating their performance. By enabling the possibility of verifying the performance of a real hardware in a simulated grid, PHIL testing combines the advantages of experimental validations with the flexibility and of simulations. This allows to test new system components in various grid conditions, including abnormal situations such as extreme frequency deviations, which are not possible or safe to carry out in a field test, due to the safe operational limits of real power systems. Moreover, since the real hardware is tested in PHIL testing, it can reveal existing issues, which are difficult to detect using only numerical simulations [18].

Recently, the PHIL testing of lithium-ion battery energy storage systems [19]–[21], supercapacitors [22], and hybrid battery-supercapacitor storage systems [23] have been reported in literature for different applications, including frequency support. However, to our knowledge, the PHIL testing of a FESS in AC grids, in particular for the application of frequency support has not yet been conducted. There are only limited experiments carried out using a FESS in shipboard DC power systems, where the FESS is supporting pulsed power loads [24], [25].

This work aims to conduct a comprehensive study on the integration of a FESS in low voltage distribution grids and validating its performance by means of PHIL testing. In particular, the following objectives has been defined:

- Finding the most suitable location for the installation of a FESS in a local grid in southern Germany in collaboration with a distribution system operator
- Deriving the required characteristics (e.g., rated power and capacity) of a FESS from collected measurements data at several low voltage distribution grids
- Dynamic modeling, real-time simulation, and model validation of a FESS
- Establishing a laboratory setup for PHIL testing of a FESS, and developing benchmark grids and test cases in a real-time simulation environment for testing the FESS in different scenarios
- Conducting PHIL testing of a high-speed FESS in several scenarios, evaluating its transient performance, and ensuring its compliance with the latest grid code requirements in Germany
- Investigating new control designs for the FESS for enhanced frequency services in low-inertia power systems, and experimental validation of the new controllers on a real FESS using rapid control prototyping and PHIL testing

With the aim of addressing the aforementioned objectives, the content of this thesis is organized as follows. In Chapter 2, the FESS is introduced and its main characteristics in comparison to other storage technologies are discussed. The structure of the FESS, its main components, and its major applications in power systems are presented. Here, the latest developments in using a FESS for different applications are reported, including novel control designs and advancements in system components.

Chapter 3 gives a short introduction to the applications of real-time simulations in power systems, including the PHIL testing of distributed energy resources and energy storage systems. The challenges in real-time simulation of low voltage distribution grids, which is an important prerequisite for the PHIL testing of any component in a low voltage grid, are discussed, followed by a summary of the available solutions and techniques to tackle these challenges. Next, a step-by-step guideline for conducting PHIL testing of new components is provided. Here, the recent efforts in PHIL testing of energy storage systems are also presented. Lastly, the 1 MVA PHIL testing facility at KIT's EnergyLab 2.0 and its main components are briefly described.

The problem of allocation and sizing of a FESS in low voltage distribution grids is addressed in Chapter 4. Firstly, the optimal placement of a FESS is investigated by locating the grid connection points with the highest voltage sensitivity to active power changes. After discussing the limitations of the classical voltage sensitivity calculations, a data-driven approach based on the concept of mutual information [26] is introduced. The allocations of a FESS in a real grid in southern Germany is presented as a use case of the proposed approach. The second part of this chapter deals with the problem of sizing of storage systems using historical measurement data. It introduces a new sizing methodology, which uses reoccurring daily consumption patterns detected by the motif discovery algorithm [27], rather than using an arbitrary day, as commonly found in literature. The proposed method is applied for deriving the characteristics of two types of energy storage systems, including a FESS.

Prior to conducting the PHIL tests of any component, having a model of the hardware under test can be significantly beneficial for designing safe and successful experiments. Therefore, a dynamic model of a high-speed FESS is developed in Chapter 5, which consists of the model of each individual component of the FESS, including the permanent magnet synchronous machine, the voltage source converters, and their controllers. The model is simulated in real-time, which allows testing the FESS model in the same grid models that are later used for the PHIL testing of a real FESS.

In Chapter 6, the performance of a 60 kW/3.6 kWh high-speed FESS with a maximum rotational speed of 45,000 rpm is evaluated by means of PHIL testing in several frequency deviations scenarios. The response of the FESS to frequency disturbances, including the major frequency incident of August 9, 2019 in the power system of the UK, and its compliance with the latest grid code requirements in Germany are investigated. The description of the PHIL setup assembled and used for the testing the FESS is also provided. The results of the PHIL tests in all investigated scenarios are also used to validate the model of the FESS developed in Chapter 5.

As discussed, the prompt response of a FESS can help maintaining the grid frequency, in particular in low-inertia systems, where any imbalance in the generation and the demand can lead to rapid changes in the grid frequency. Therefore, Chapter 7 proposes the use of a FESS for inertia emulation, and a new adaptive controller for this purpose is introduced. The performance of the proposed design is first evaluated by means of simulations in a low-inertia microgrid, and compared with several previously reported adaptive controllers for inertia support. Next, the controller is implemented on the 60 kW high-speed FESS using the concept of rapid control prototyping, and the performance of the FESS with the proposed controller is verified using PHIL testing of the FESS in a low voltage microgrid in different scenarios.

Finally, Chapter 8 summarizes all the findings, and illustrates the potential applications of the proposed methods, and the future possible research in these areas.

2 The Role of Flywheel Energy Storage Systems in Power Systems

In this chapter, the FESS, its characteristics, and its main applications in power systems are described. For each application, including frequency regulation and power smoothing, the current state-of-the-art together with several real-world implementations are presented. Moreover, the structural differences between the main two types of FESS, i.e., the low-speed and the high-speed FESS, are provided. Lastly, a comparison between various technologies used for the main components of a FESS is given.

2.1 Introduction to Flywheel Energy Storage Systems (FESS)

A FESS is a mechanical energy storage system, which stores energy in the form of kinetic energy by accelerating a large rotating mass with a high inertia, i.e., the flywheel. The stored energy can be extracted from the FESS by deaccelerating the high-inertia rotor. While the concept of a FESS is considered among the oldest storage technologies, recent developments in high tensile strength materials and advanced magnetic bearings have made it possible to increase the energy density and efficiency (above 85 % [28]) of this technology, significantly. The configuration of a conventional FESS is illustrated in Figure 2.1. The term FESS refers to the entire system consisting of a flywheel or a high-inertia rotor, an electrical machine, back-to-back bi-directional power converters with a common DC link, and auxiliary components required for running the system.

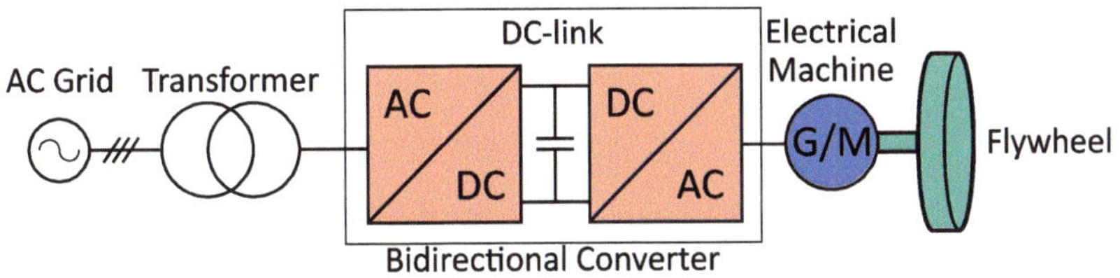

Figure 2.1: Configuration of a conventional Flywheel Energy Storage System (FESS).

The advantages and disadvantages of the FESS in comparison to other storage technologies are listed in Table 2.1. A FESS has several important advantages, including high power density, fast response time and cycling characteristic, high number of cycles, long lifetime, high controllability, insensitivity to ambient temperature, and environmentally benignity. Moreover, the capacity of a FESS remains consistent throughout its entire lifetime, independent of the way it has been charged or discharged and the ambient temperature. In terms of the specific power and energy, the flywheel technology bridges the gap between the supercapacitors and lithium-ion batteries, as shown in Figure 2.2. While a FESS can have the power capabilities of a supercapacitor, it can have a higher specific energy, making longer discharge times possible. Moreover, a FESS has significantly longer lifetime and a greater number of cycles in comparison to lithium-ion batteries.

Table 2.1: Advantages and disadvantages of a FESS in comparison to other storage technologies.

Advantages	Disadvantages
+ High power density (up to 2 kW per liter)	- High capital costs
+ Extremely fast cycling characteristics	- Low energy density
+ Fast response time (in the range of milliseconds)	- High self-discharge rate
+ Long lifetime (in the range of 20 years or 200,000 cycles)	
+ Constant capacity throughout its lifetime	
+ Not sensitive to the ambient temperature	
+ Environmentally friendly	

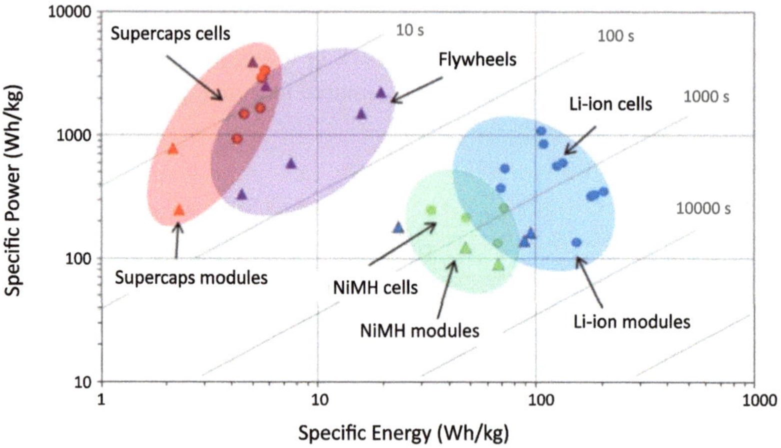

Figure 2.2: Comparison of the specific power and energy of the FESS with supercapacitors and lithium-ion (Li-ion) batteries (adopted from [29]).

Limited storage capacity, high self-discharge rate, and high capital costs are considered as the main barriers towards a widespread use of FESS [30]. Higher energy densities and lower self-discharge rates are achieved by the use of composite fiber materials and advanced magnetic bearings to reach higher rotational speeds, but these technologies increase the production costs as well [13]. In addition, the capacity of a FESS can only be increased to a certain extent, due to the limited tensile strength of commercially available rotor materials, which constraints the maximum rotational speed of the rotor (see subsection 2.3.1 for more details). At present, the investment costs of a FESS are relatively high in comparison to several other storage technologies, which could be due to the limited number of flywheel manufacturers worldwide. A list of current manufacturers of a FESS with the characteristics of their product is provided in Table 2.2. With growing interest in the use of a FESS for different applications, the number of manufacturers can increase, which can potentially reduce the capital costs of the FESS. It is estimated that the capital costs of this storage technology would fall by 35 % by 2030 according to a report by the International Renewable Energy Agency (IRENA) [2]. Safety concerns in the case of system failures, are also of crucial importance when using flywheels, but these risks are reduced when light composite fiber materials are used for the rotor [31] (see subsection 2.3.5). While lithium-ion batteries will remain the dominant storage technology for the new storage installations worldwide [32], for applications which require short-term storage of high amounts of power with a high number of cycles, other technologies such as FESS can be more suitable [33].

Table 2.2: A comparison of the characteristics of several commercial FESS [34].

Company	Rated Power	Rated Capacity	Maximum Speed	Lifetime	Main Application	Country
Beacon Power	100 kW	25 kWh	16,000 rpm	175,000 cycles	Grid services	USA
Piller	1-3 MW	10-17 kWh	3,400 rpm	> 20 years	UPS	Germany
Stornetic	60 kW	3.6 kWh	45,000 rpm	> 20 years	Grid services	Germany
Gyrotricity	100 kW	5 kWh	20,000 rpm	25 years	Electric vehicle charging	UK
Amber kinetics	8 kW	32 kWh	10,000 rpm	11,000 cycles	Grid services	USA
Active Power	225 kW	up to 2.6 kWh	7,700 rpm	20 years	UPS	USA
Vycon	450 kW	1.7 kWh	36,750 rpm	20 years	UPS, grid services, mobility	USA
Adaptive Balancing Power	100-500 kW	5-12 kWh	N/A	1 M. cycles	Grid services, mobility, industry	Germany
Kinetic Traction Systems	333 kW	1.5 kWh	36,000 rpm	10 M. cycles	UPS, railway, grid services	USA

2.2 Applications of FESS in Power Systems

A FESS can be used for a variety of applications in different sectors, depending on its ratings and its type (see section 2.3). Table 2.3 lists the applications of this storage technology in different sectors and areas. In the field of power systems, a FESS is most commonly used for frequency regulation and smoothing out short-term power fluctuations of wind generation units. In weakly interconnected power systems, such as in microgrids, these two applications can be intertwined, as mitigating the power variability of renewables simultaneously leads to a more stable frequency. In addition, a FESS can also be used for improving the system transient stability, voltage regulation, spinning reserve, black start, reliability enhancement, fault ride-through support, and unbalanced load compensation. However, these applications are less common in comparison to the other applications and are less specific to the flywheel technology. The use of a FESS can be particularly beneficial, where a significant number of high-power short-duration cycles ranging from seconds to several minutes are required.

As seen in Table 2.3, FESS are also used in several other sectors, including in the transportation sector [35], for pulsed power applications in industry [36], and in space satellites [37]. These applications are out of the scope of this thesis, and therefore, not discussed further. In the following, the applications of a FESS in power systems are described briefly together with a review of the latest works in each area.

Table 2.3: Applications of FESS in different sectors and fields.

Sector	Applications
Power Systems	Frequency regulation, power smoothing, voltage regulation, improving transient stability, spinning reserve, ride through support, unbalanced load compensation, and as Uninterruptable Power Supply (UPS)
Transportation and mobility	Vehicles, trains and subways, lifts, etc.
Industry	Mining, cranes, lasers, linear motors, and other pulsed power applications
Research	Pulsed power for fusion research facilities (e.g., Joint European Torus, ASDEX)
Space	Satellites attitude control, energy buffer

2.2.1 Frequency Regulation

As discussed, a FESS is a well-suited solution for supporting the grid frequency during imbalances between generation and demand, due to the high power density, quick response time, high reliability, long lifetime, and high number of cycles with no degradation.

A single flywheel unit can be used in microgrids to support the frequency, or tens to hundreds of units can be connected in parallel in a modular configuration to build a utility-scale system. For instance, the company Beacon Power has built and successfully commissioned two 20 MW plants, each with 200 high-speed flywheels in Stephentown, New York, and in Hazel, Pennsylvania. These plants provide fast frequency regulation services to the grid. For the state of New York, the flywheels provide approximately 10% of the whole state frequency regulation requirement. The field data have confirmed the advantages of these systems in terms of costs, emissions, and operational constraints [38].

The combination of the flywheel technology with a battery for providing frequency support has also gained attention over the last few years. Recently, a 9 MW hybrid flywheel-battery energy storage system has been commissioned in the Netherlands to be used for enhanced frequency control, where the flywheel increases the power capabilities of the battery [39]. A smaller flywheel-battery hybrid storage system with the rating of 480 kW was successfully tested in Ireland, where the flywheel responds faster with a higher ramp rate to frequency deviations in comparison to the battery, by design [7]. Similarly, in [40], authors proposed the use of a 50 MW flywheel-based plant for the northern Chilean interconnected system, which can quickly provide the required power for primary frequency control and improving the transient stability of the power system.

Applications of a FESS in a microgrid, in particular for regulating the frequency, have drawn a significant attention over the last 20 years, both in academia and industry. A considerable number of projects have been conducted globally in this area. Since a FESS can simultaneously supply the microgrid with both active and reactive power with a short response time, it can improve the power quality and the dynamic security of the microgrid, significantly. For example, a 500 kW/5 kWh FESS was designed and installed in the Portuguese island of Flores. It was shown that with the help of the FESS, a stable frequency is feasible, even with a high penetration of renewable energy sources [41].

In literature, several new control structures for a FESS for the application of frequency regulation have been proposed. Authors in [42] propose an enhanced frequency control for a low-speed FESS with the aim of reducing the frequency and the DC-link voltage deviations. In [43], a coordinated control strategy between a FESS and a battery energy storage system has been suggested for a more effective droop-based frequency control in microgrids. A fuzzy logic controller for controlling multiple storage systems including a FESS has been introduced in [44], which maintains the state of charge of each storage system in a certain range, while providing enhanced frequency services. Recently, the concept of inertia emulation using a FESS has been introduced [45], and a virtual synchronous generator based on a FESS has been suggested [46]. More information on these works and inertia emulation using a FESS are provided later in Chapter 7.

2.2.2 Power Smoothing

The advantages of using a FESS is often investigated in the presence of wind generation units, where the FESS can mitigate the high variability of the power generated by wind turbines. This is particularly important for integrating wind generating units into weak or isolated microgrids. With a high number of cycles and no capacity loss, a FESS is a suitable candidate for power smoothing applications, where a continuous operation of the energy storage systems is often required.

Several real-world projects including the FESS in the Glencore Raglan Mine in Quebec has been reported, where the FESS smooths the power produced by a 3 MW wind turbine. Similarly in the Saint Paul Island in Alaska, a FESS is used to reduce wind power fluctuations and the fuel consumption of two diesel generators [47].

In literature, numerous novel control methods have been proposed for the use of a FESS for smoothing the power generated by wind turbines. Recently, an intelligent controller based on the adaptive wavelet fuzzy neural networks has been introduced to mitigate the intermittency in wind generation using a FESS, and improve the transient stability of the power system [48]. The authors in [49] have suggested an energy function-based optimal control approach for reducing the effect of wind power fluctuations on the power system. Here, it has also been shown that the FESS can also improve the low voltage ride-through capability of the wind turbine by operating as a static compensator (STATCOM) during low voltage incidents. In another study, the flux-oriented control and the direct torque control have been compared for a low-speed FESS, and it has been argued that the direct torque control is a better solution for smoothing wind power generation using a FESS [50]. The authors in [51] have presented a control algorithm for a FESS to simultaneously achieve mitigation of power fluctuation and dynamic stability enhancement for an offshore wind farm. Moreover, a supervisory control unit with short-time wind speed prediction has been suggested to be implemented on a wind farm equipped with a FESS, which has proven to be effective for reducing power fluctuations and the size of the required FESS [52].

By compensating for the short-term volatility of renewable energy resources, a FESS can also reduce the variability of the power output of other distributed energy resources and storage technologies. For instance, a FESS can reduce the power variability of a diesel generator, and therefore, reduce its fuel consumption and emissions, improve its efficiency and lifetime, and decrease the number of required startups and shutdowns, when the generator is used as a backup system [41], [53]. In Nullagine and Marble Bar in western Australia, a 500 kW FESS from ABB in combination with large PV systems have managed to reduce the fuel consumption of the installed diesel generators by 400,000 liters per year and save 1100 tons of greenhouse gas emissions annually [54].

When used together with a battery in a hybrid configuration, a FESS can significantly help preserving the lifetime and capacity of the battery [55], [56]. It has been demonstrated that a battery-flywheel hybrid system outperforms a battery-supercapacitor combination in terms of reducing the peak current drawn from the battery and the overall system efficiency [57]. In Australia, a 2 MW FESS is currently being used to relief a 3 MW battery from wearing faster for mitigating wind power fluctuations [58].

2.2.3 Voltage Regulation

A FESS, like any converter-based system can also provide reactive power for regulating the grid voltage at its connection point and operate similar to a static compensator (STATCOM). However, in low voltage distribution grids with a low X/R ratio, active power provided by the flywheel can be more effective in regulating the grid voltage, in comparison to reactive power (see subsection 4.1.1).

Dynamic voltage compensation on distribution feeders using a FESS has been studied in [59], where it has been shown that a series-connected FESS is more effective for voltage regulation than a FESS connected in parallel to the grid. A series-connected FESS equipped with a matrix converter has been designed and built in [60] for compensating voltage sags, and in [61], a predictive optimal controller has been suggested for such a system. It is also proposed to use a FESS for unbalanced voltage drop compensation and reducing the effect of asymmetric loads on the grid [62].

2.2.4 Other Applications in Power Systems

As listed in Table 2.3, a FESS can be used for several other applications in power system as well. For instance, it has been suggested to use a FESS for enhancing the fault ride-through capability of an HVDC connection to a wind farm during AC faults [63]. The use of a FESS with a self-organizing fuzzy neural network controller to increase the power transfer capability and transient stability of the power system has been suggested by the authors in [64]. More recently, it has been suggested to use a FESS to limit the initial high power demand of electric vehicle fast chargers from the grid at the moment of connection [65].

2.3 Components of FESS

Flywheel energy storage systems can be grouped into two main categories, i.e., the low-speed (less than 10^4 rpm) and high-speed ($10^4 - 10^5$ rpm) FESS, which are very different in their structure and design, and are often used for different applications. A comparison of the structural differences between the two types of the FESS is presented in Table 2.4. Low-speed FESS are the cheaper alternative, which are suitable when a discharge time of several seconds is adequate and a high self-discharge rate is not a concern. High-speed FESS use magnetic bearings, composite fiber rotor, and a vacuum enclosure to reach higher rotational speeds and storage capacities and a significantly reduced self-discharge rate. However, they are significantly more expensive than the low-speed systems. While low-speed FESS are considered a matured technology, high-speed FESS have been only recently commercialized. Figure 2.3 shows the structure of a commercial high-speed FESS.

In the following subsections, important components of a FESS and the different technologies used for each component are discussed.

Table 2.4: Comparison of the main two types of FESS [66], [67].

Component/Characteristic	Low-speed FESS	High-Speed FESS
Flywheel material	Steel	Composite Materials: Glass or carbon fiber
Electrical machines type	Asynchronous, permanent magnet synchronous or reluctance machines	Permanent magnet synchronous or reluctance machines
Integration of machine/flywheel	No integration or partial integration	Full or partial integration
Confinement atmosphere	Partial vacuum or light gas (Helium)	Absolute vacuum
Enclosure weight	2 × Flywheel weight	½ × Flywheel weight
Bearings	Mechanical or hybrid (mechanical and magnetic)	Only magnetic
Relative capital cost	1	5
Main applications	Short-term and medium power applications	Power quality improvement, ride-through support, power service in traction and the aerospace industry
Specific energy	~ 5Wh/kg	~ 100Wh/kg
Technology maturity	Commercialized, mature technology	Early commercialization

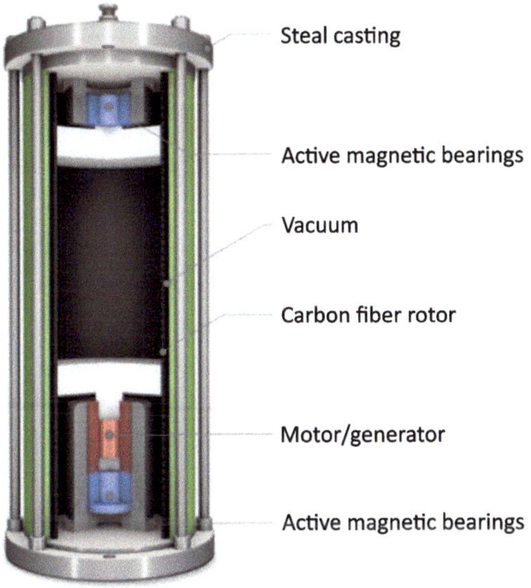

Figure 2.3: Structure of the container holding the rotor and the electrical machine in a high-speed FESS (Picture courtesy: Stornetic GmbH [68], used with permission).

2.3.1 Rotor

The energy of a FESS (E_{FW}) depends only on the rotor inertia (J_f) and its rotational speed (ω_m) according to Eq. (2.1).

$$E_{FW} = \frac{1}{2} J_f \omega_m^2. \tag{2.1}$$

As seen in Eq. (2.1), while the capacity of a FESS can also be expanded by increasing the rotor inertia, reaching higher rotational speeds has a greater impact on the storage capacity. The maximum rotational speed is limited by the strength of the rotor material to withstand the centrifugal forces affecting the rotor at high rotational speeds [28]. This is referred to as the material tensile strength. The maximum stress that the rotor material experiences is proportional to the material density and the square of the rotor speed at the rotor's outer surface [69]. The maximum stress has to be below the tensile strengths of the used material by a safe margin. If the material tensile strength is represented by σ, the energy density (e_v), and specific energy (e_m) is calculated using [66]

$$e_v = K\sigma, \tag{2.2}$$

and

$$e_m = \frac{K\sigma}{\rho}. \tag{2.3}$$

In Eq. (2.2) and (2.3), ρ is the density of the rotor material and K is called the shape factor, which depends on the rotor geometry, and it is a measure of effective material utilization [31]. Equation (2.3) indicates that in order to achieve a higher capacity per unit of mass, a material with a high tensile strength and a low density is required. These are the characteristics of composite fiber materials with glass or carbon, as seen in Table 2.5. Therefore, high-speed FESS use composite fiber materials, which make it possible to reach rotational speeds up to 100,000 rpm. Low-speed FESS commonly use variations of steel such as carbon steel [70], which are significantly cheaper than the composite fiber materials. In a low-speed FESS, since the rotational speed of the rotor is limited by the tensile strengths of steel, the capacity is often increased by using larger bulkier rotors (see the two-ton rotor in [71], as an example).

According to Eq. (2.2) and (2.3), the shape factor K also influences the energy per unit of volume and mass. The shape factor depends only on the geometry of the rotor. The rotor geometry can also be optimized to reach higher energy densities, as shown in [72]. However, in practice, solid and hollow cylinders are commonly used for low-speed and high-speed FESS, respectively.

Table 2.5: Characteristics of several materials used for the rotor in a FESS [28], [66].

Material	ρ (kg/m^3)	σ (MPa)	e_m (MJ/kg)
Aluminum	2700	500	0.093
Steel	7700	1520	0.19
Glass E/Epoxy	2000	1000	0.25
Carbon AS4C	1510	1650	1.1
Carbon T1000	1520	1950	1.28

While the energy of a FESS depends on the rotor material and geometry, its maximum power depends on the ratings of the electrical machine and power converters, which are described in the following subsections.

2.3.2 Electrical Machine

The high-inertia rotor of a FESS can be coupled to various types of electrical machines, including different designs of synchronous machines, asynchronous machines, and reluctance machines. Each machine type and design has its own advantages and disadvantages. Table 2.6 presents a comparison between the important characteristics of the most commonly used electrical machines in a FESS.

Asynchronous machines are often used in a low-speed FESS. They have a robust construction, high torque, high reliability, and low costs. However, due the presence of copper losses in the rotor, they cannot be used in the vacuum enclosure of a high-speed FESS, where convection cooling is not available. Therefore, a Permanent Magnet Synchronous Machine (PMSM) is often used in a high-speed FESS, which has no field winding and consequently, no copper losses on the rotor. Moreover, PMSM have high specific power, high efficiency, simple and robust structure, and excellent dynamic performance [73]. However, they carry the risk of demagnetization under high temperatures [74]. They also have higher costs and lower tensile strength comparing to asynchronous machines.

Other types and designs of electrical machines such as switched reluctance machines [75], homopolar machines [76], and axial-flux machines [77] have also been used in a FESS. However, they are rarely used in practice in commercial systems.

Table 2.6: Comparison of different types of electrical machines used in FESS [53], [78].

Machine type/ Characteristic	Asynchronous	Reluctance	Permanent magnet synchronous
Power	High	Medium & Low	Medium & Low
Specific power	Medium (0.7 kW/kg)	Medium (0.7 kW/kg)	High (1.2 kW/kg)
Rotor losses	Copper & iron	Iron (Slots)	None
Spinning losses	Removable by annulling flux	Removable by annulling flux	Non-removable, static flux
Efficiency	High (93.4 %)	High (93 %)	Very High (95.2 %)
Power density	1.8 l/kW	2.6 l/kW	2.3 l/kW
Tensile strength	Medium	Medium	Low
Torque ripple	Medium (7.3 %)	High (24 %)	Medium (10 %)
Max./base speed	Medium (>3)	High (>4)	Low (<2)
Demagnetization	No	No	Yes
Cost	Low (22 €/kW)	Low (24 €/kW)	High (38 €/kW)

2.3.3 Power Converters

The power converters control the power exchange between the high-inertia rotor and the grid. A FESS is most commonly connected to the AC grid via back-to-back converters with a common DC link [79], as shown in Figure 2.1. These converters convert the output of the electrical machine with a variable frequency to the fixed frequency required by the grid. For some specific applications of flywheels, such as in wind turbines [80], in uninterruptable power supplies [81], or in electric cranes [82], the flywheel can also be coupled through the DC link of an existing converter using a separate DC-AC converter, as depicted in Figure 2.4. In some cases, an extra bi-directional DC-DC converter may be required to reach the required voltage level in such configurations [83].

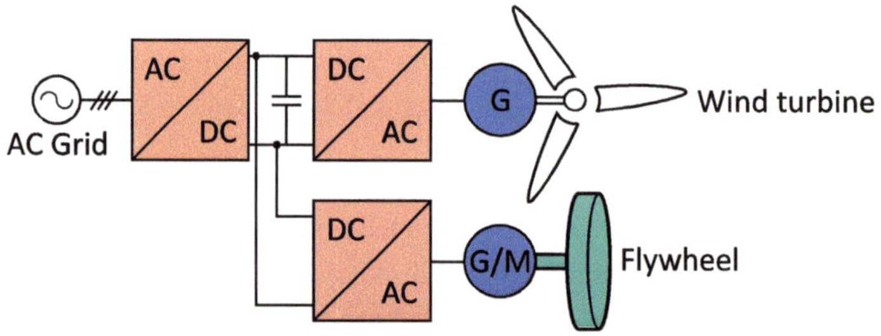

Figure 2.4: Flywheel connected to the common DC link of a wind generator for power smoothing applications.

Power converters in a FESS are often conventional Voltage Source Converters (VSC), which are controlled using Pulse-Width Modulation (PWM). Different types of switches (MOSFETs, IGBTs, etc.) can be used, depending on the blocking voltage and current requirements and the switching frequency. Recently, the use of silicon carbide-based MOSFETs has also been investigated for a FESS [84]. Higher switching frequencies in the machine-side converter leads to a reduction in the current and torque ripples of the electrical machine. For the grid-side converter, this leads to decreased harmonics injected to the grid and a smaller grid filter. But this increases the switching losses as well (except when silicon carbide switches are used). Due to the low stator inductance of the electrical machines often used in a high-speed FESS, an LC filter is often required to connect the machine-side converter to the electrical machine, in order to reduce the current ripples and the machine losses [28], [85].

In terms of the converter topology, two-level converters are most commonly used in practice in a FESS. The use of a diode-clamped three-level topology, also referred to as the neutral-point clamped converters, is also suggested for a FESS. This configuration leads to less harmonic distortion, lower machine losses, additional flexibility of the output voltage, higher efficiency, and faster dynamic response, in comparison to a conventional two-level converters, under the same switching frequency [86]. A matrix converter is another alternative suggested to be used in a FESS [60], which does not have a DC-link capacitor in between, and directly converts AC to AC. This reduces the volume and weight of the converters, significantly. However, matrix converters have high harmonic distortions and more complex control and protection systems. The authors in [87] have presented a FESS with a Z-source converter and claimed that this configuration leads to a higher efficiency. Other novel topologies and switching schemes have been suggested to improve the converter efficiencies in a FESS, such as in [88], [89]. However, such systems are rarely used in industry.

2.3.4 Bearings

The flywheel bearings have to support the weight of the rotor, keep it in position, damp out mechanical vibrations (referred to as stiffness), and allow the free rotation of the rotor with minimum losses. Table 2.7 presents the advantages and disadvantages of various bearing technologies used in a FESS.

Traditionally, a low-speed FESS uses mechanical bearings, which have high friction losses and lubrication and maintenance requirements. The friction losses lead to a high self-discharge rate and a low lifetime in such systems. Magnetic bearings are used as the alternative solution, in which the rotor is suspended using a magnetic field, often in a vacuum enclosure, eliminating friction losses and wearing. Therefore, they require much less maintenance, and significantly lower self-discharge rates and higher rotational speeds can be achieved. Magnetic bearings lack friction losses, but other losses resulting from the geometry and variance in a magnetic field, such as eddy current losses, leakage fluxes (stray flux paths), and hysteresis losses are present. Auxiliary mechanical bearings are still required for the initial rotation of the flywheel, or in a case of a magnetic bearings failure or an overload [67].

Magnetic bearings are categorized into active and passive bearings. In active magnetic bearings, the magnetic field is controlled using a coil and a control system with respect to the position and the movement of the rotor. This provides a high stiffness and damping, and results in a highly reliable system with a long lifetime and low maintenance [28]. However, the extra power and the complex control system required for operating such systems are considered as the disadvantages of this type of bearing. They increase the system costs as well.

Passive bearings include the use of permanent magnets. However, permanent magnets should be used with another type of bearing, as they are inherently unstable [53]. The use of High-temperature Superconductors (HTS) in combination with permanent magnets as flywheel bearings leads to less intrinsic losses, compared to any other types of magnetic bearings [90]. Figure 2.5 shows the structure of a flywheel with HTS bearings. The HTS also resists the movement of the rotor due to the pinning effect in HTS materials, which prevents the motion of flux lines [91]. However, HTS bearings have less levitation force, lower stiffness, and lower critical frequencies, and require a cryogenic cooling system. Therefore, hybrid configurations, combining different types of magnetic bearing were proposed. For instance, HTS bearings can be used to either support the weight of rotor or to stabilize the rotor, and active magnetic bearings can be used to improve the stiffness and the damping [92]. Several demonstrators of a FESS with HTS bearings have been built and successfully tested, including in [90], [93], [94]. However, there is no commercial FESS currently available in the market, which uses HTS bearings.

Table 2.7: Comparison of different bearing technologies used in FESS.

Bearing Technology	**Advantages**	**Disadvantages**
Mechanical	Low capital costs, matured technology, high levitation force, high critical frequencies	High losses, short lifetime, required lubrication, high maintenance, most sensitive to rotor imbalances
Permanent Magnetic	No power consumption, low losses	Inherently unstable, risk of demagnetization, low tensile strength, uncontrollable
Active Magnetic	Controllable magnetic field, high stiffness and damping, high reliability	Complex systems, large consumption of power
HTS Magnetic	Auto-stabilizing behavior, less losses than other magnetic bearings, least sensitive to rotor imbalances	Low levitation force, low stiffness, low critical frequencies, requires a cryogenic cooling system

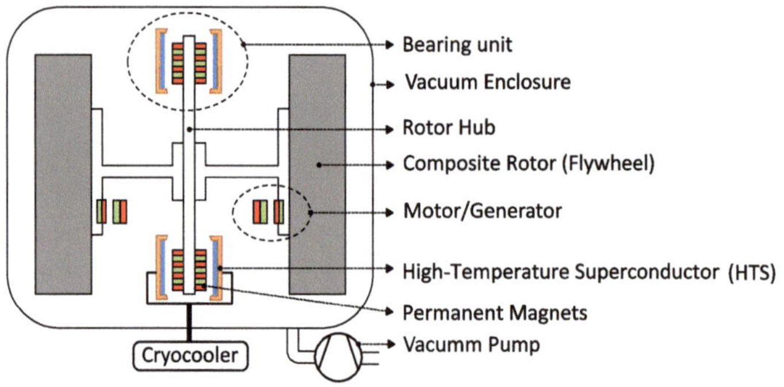

Figure 2.5: Structure of a FESS with of High-temperature Superconductors (HTS) bearings.

Bearing losses have to be calculated when designing a FESS. Losses in a FESS also include the losses in the electrical machine, power converters, and AC filters. Moreover, the power required for supplying the active magnetic bearings, vacuum and cooling pumps, control systems, and other auxiliary system required for running the FESS can also be considered as the system losses.

2.3.5 Enclosure

The major task of the flywheel enclosure is to contain the fragments of the rotor in case of a disastrous system failure at high rotational speeds and to dissipate its energy content. Low-speed FESS require thick heavy enclosures, as a steel rotor can be forcefully dismantled into only few pieces with translational movements [95]. On the contrary, composite rotors used in a high-speed FESS tend to disintegrate into numerous small pieces with rotational movements, resulting in energy dissipation caused by friction. As a results of these differences, it has been reported that for the same energy content of the rotor, the enclosure weight of a high-speed FESS needs to be only half of the rotor weight, while for low-speed FESS, it should be the double [53]. Experimental crash tests are necessary in order to ensure the system safety in case of failures at high speeds.

In a high-speed FESS, the enclosure is also used to provide a vacuum surrounding the rotor, minimizing the air friction and system losses. In low-speed systems, the vacuum can be partial or a low-pressure light gas such as helium is used, since the friction torque is proportional to the density and pressure of the gas surrounding the rotor [53].

2.4 Summary

In this chapter the FESS was introduced along with its main characteristics and applications. As discussed, a FESS has a high power density, quick response time, and high number of cycles with a constant capacity through its lifetime, independent of the way it has been used. Considering these characteristics, the FESS is well suited for applications such as frequency regulation and smoothing out short-term power variations. For each application, the latest developments in academia and industry were reported. Furthermore, the structural differences between the low-speed and the high-speed FESS were presented, where it was shown that how a high-speed FESS can reach higher energy densities and lower losses with the use of light composite fiber materials and advanced magnetic bearings. Therefore, the high-speed FESS is a more promising alternative for power system applications. Lastly, a detailed overview and comparison of available technical solutions together with newly introduced alternatives for each component of the FESS were provided.

3 Real-time Simulation and Power Hardware-in-the-Loop Testing for Grid Integration Studies

In this chapter, real-time simulations and their applications in power systems are introduced. The challenges in real-time simulation of low voltage distribution grids are discussed, and the solutions and the techniques to reduce the model execution time of low voltage grid models are presented. Next, the concept of Power Hardware-in-the-Loop (PHIL) testing is described as a powerful tool for grid integration studies of new grid components, such as energy storage systems. A step-by-step guideline for conducting PHIL testing is presented. The main concerns associated with PHIL testing, including instability issues are also discussed. Lastly, the 1 MVA PHIL testing facility at KIT's EnergyLab 2.0 is described.

3.1 Challenges of Real-time Simulations of Low Voltage Distribution Grids

In real-time simulations, by definition, the status of the simulated model must be updated at every small simulation time step in order for the model to accurately resemble the behavior of its physical counterpart [96]. In the case of power systems, this requires that the complex and large differential equations of the electromagnetic transient model of the corresponding power system to be solved within a predetermined simulation time step. The simulation time step for power system transient simulations can be as low as tens of microseconds, depending on the application and the power system dynamics to be simulated [97]. All calculations have to be finished within the given time step at all times. This is known as the hard real-time constraint. On the contrary, in non-real-time power system simulations, there are no constraints on the time available to solve the model equations. Also, variable time-step solvers can also be used, reducing or increasing the simulation time step, when necessary. This is not an option for real-time simulations, which can only apply fixed-step solvers [96]. If the real-time constraint is violated at any time during the simulation, an overrun has occurred, as shown in Figure 3.1. Overruns can lead to instability and incorrect results, and therefore, should be avoided [98]. Overruns can be resolved by increasing the simulation time step, simplifying the models, or distributing the computational effort on several cores, when possible, without losing the model accuracy or the real-time granularity.

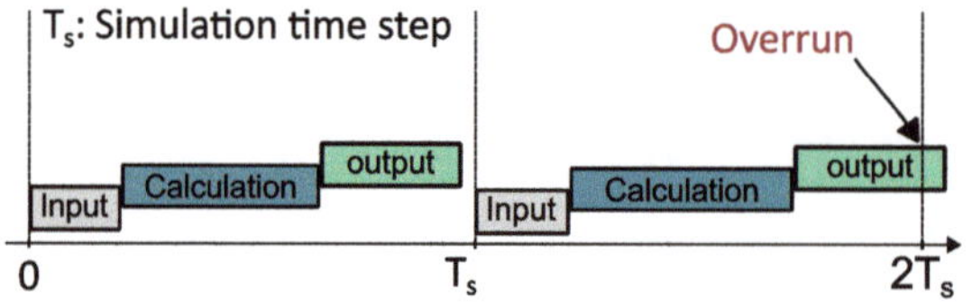

Figure 3.1: The hard real-time constraint and the definition of an overrun.

Real-time simulations can reduce the simulation time of power systems, in particular for large complex grids with a high penetration level of distributed energy resources. It enables modifying the model parameters, while the model is running, without the need for recompiling and rerunning the model. Since real-time simulation models run at the same rate as the actual system, they are also referred to as a *Digital Twin* of the simulated power system [99]. However, the main application of real-time simulations is its capability for real-time data exchange with a physical component. This concept, which is referred to as Hardware-in-the-Loop (HIL) and Power Hardware-in-the-Loop (PHIL) testing is later discussed in detail in this chapter in section 3.2. A comprehensive overview of the possible applications of real-time simulation in power systems is provided in [100].

There are several commercial and non-commercial real-time simulators available, which are tailored for power system simulations. A detailed comparison of the available real-time simulators, in terms of their components, structure, design, and capabilities are given in [97]. These real-time simulation platforms were initially designed for the simulation of large high voltage transmission grids, with the aim of testing protection systems or new components such as static var compensators [100]. Over time, new challenges rose in low voltage distribution grids, with the introduction of converter-interfaced components, such as photovoltaic systems, energy storage systems, and electric vehicle chargers. These new challenges called for innovative solutions, which required new testing methods such as Hardware-in-the-Loop (HIL) and Power Hardware-in-the-Loop (PHIL) testing. This led to a clear interest from both industry and academia to validate these technologies using these methods. For connecting a real hardware to a simulated grid, running the grid model in real-time is mandatory. However, this can be challenging in case of some low voltage distribution grids. For real-time simulation of transmission grids, the propagation delays in long transmission lines are often used for model decoupling and parallelization. However, such methods cannot be applied to low voltage grids, where lines and cables are much shorter [101]. Moreover, many converter-interfaced distributed energy resources are often connected at low voltage grids. These components can increase the model computational time significantly, if modelled in detail. These challenges and the available solutions for tackling them are discussed in detail in the following subsections.

3.1.1 Lack of Decoupling Points for Parallel Computing

Most computers have a multi-core design nowdays. To be able to use this advantage for real-time simulations, the model must have the capability to split its solution matrix and solve it on parallel cores, without losing its accuracy. Real-time simulation software can take advantage of the natural propagation delays in transmission lines for model decoupling, when these delays are greater than the simulation time step. By assuming a simulation time step of 50 μs for generic power system simulations [97], and 300,000 km/s as the wave propagation speed through lines (speed of light), this requires an overhead line to be longer than approximately 15 km in order to have the desired delay for model decoupling [102]. For underground cables, the minimum line length for model decoupling is reduced to 10 km, by assuming a propagation speed of 200,000 km/s for cables. If such relatively long lines are present in the model, the two grid areas on either side of the line can be solved independently on multiple cores using the Bergeron traveling wave line model. However, in low voltage distribution grids, lines and cables are usually much shorter [101] and other solutions have to be found.

Researchers have proposed methods to go around this issue by, for instance, introducing artificial delays into the system with the same length of the simulation time step [103]. However, such delays compromise the model accuracy, significantly [104], [105]. Another solution is to add a spurious shunt capacitor after a transformer [102]. This results in a decoupling possibility at the substation. However, since such a capacitor does not exist in reality, this method is a deliberate change to the system impedance. Therefore, this method also reduces the model accuracy. It can also cause unrealistic high transient voltage values at the transformer [106].

Another important factor to be considered in real-time simulations is the type of modelling technique chosen for formulating the power system equations. The main two well-known approaches are the nodal and the state-space analysis and each power system simulator usually uses one of these techniques [107], [108]. But authors in [109] have come up with an innovative solution that combines the state-space and the nodal analysis for the simulation of electrical systems, which can solve the model equations faster for real-time applications, with the possibility for parallelization. This method is known as State-Space Nodal (SSN). The key idea of the state-space nodal approach is to introduce major nodes in the system of equations and to use these nodes as virtual decoupling points in the circuit. Then each section can be described by local state-space equations, which are linked together through the nodal analysis. As these parts are almost independent, their iterations can be parallelized on multiple cores without any artificial delays [104]. This method does not introduce any impedance change in the model, so the results are highly accurate. However, parallelization using the state-space nodal method is not as effective as other decoupling method, as parts of the calculations cannot be parallelized, and calculations are not fully independent within a single simulation time step [109].

3.1.2 Aggregation of Distributed Energy Resources

Low voltage distribution grids often have a high penetration level of converter-interfaced elements. These components have higher levels of complexity, in both structure and control, in comparison to the traditional power system elements, such as lines and transformers. Therefore, a detailed simulation of these components requires a high computation effort, in particular when they have a large number of switches at high switching frequencies. The high computation burden can even be beyond the capabilities of CPU-based real-time simulators and require running the models on an FPGA.

While for some studies a detailed converter model for distributed energy resources may not be necessary [110], for applications such as hardware-in-the-loop testing of converter controllers or assessing high-frequency harmonics, a switching or detailed converter model (see subsection 5.2.2) is required. For such simulations, the simulation sampling frequency (one over the simulation time step) should be at least 50 to 100 times larger than the pulse-width modulation frequency [111]. This means that a detailed converter model with a switching frequency of 10 kHz requires time steps of 1 μs or less. In addition, fixed-step non-iterative solvers used in real-time simulation software can hardly handle the instantaneous switching events in the converters [112]. Thus, lower simulation time steps may be required for maintaining the numerical stability. Therefore, CPU-based real-time simulators, which have a minimum simulation time step of few microseconds, may not be capable of simulating detailed models of such converters, and the use of FPGA-based simulators might be mandatory. There are several switching event compensation methods [111], [113] that can increase the required simulation time step to some extent, however, these methods also have limitations. The simulation time step in FPGA-based real-time simulation can be in the nanosecond range. In this case, co-simulation using CPU and FPGA can also be a good alternative, in which the grid model runs on the CPU, while the converter switching or detailed model runs on the FPGA with a sub-microsecond simulation time step [114]. In [101], a large distribution grid and a microgrid were simulated in real-time using this approach. However, like any multi-rate real-time simulation, the interfacing of the simulations on the CPU and the FPGA can be challenging and can cause numerical instability. This problem can be solved by adding snubber RC circuits or other solutions proposed in [115]. For enhancing the FPGA capabilities for real-time simulation of power converters with a high number of switches and high switching frequencies, several solutions have been proposed in the literature. For instance, authors in [115], have managed to decouple the power electronics circuits by using different discretization methods and combining implicit and explicit solvers.

Another solution has been proposed in [116], which allows subsets of switches to be treated independently to reduce the model computational time.

However, depending on the intention of the study, the converter models and their controllers can be simplified. For applications such as optimization, scheduling, primary frequency control, voltage stability analysis, and even DC-link voltage stability, the converter average model can be adequate [110]. An example of using average converter models for the application of frequency and voltage support using energy storage systems is later presented in Chapter 5.

There are also methods to decouple the converter model on multiple CPU cores to reduce the model execution time. This can be done using the following solutions:

- For a large DC-link capacitor, the DC-link voltage can be assumed to be constant within one simulation time step. Therefore, the DC-link can be used to decouple the converter models on multiple cores. This method is particularly effective for complex power electronics-based systems and does not introduce a noticeable error in the results [23].
- Between the converter switches and their corresponding controllers, there are communication delays, rate conversion and modulation blocks, which can all be approximated by a unit delay [24]. This delay can be used as a decoupling point between the converter controller and the rest of the system [25].

3.1.3 Presence of Asymmetry and Harmonics

Another characteristic of the low voltage grids is that the current and the voltage asymmetries and harmonics are more dominant at this voltage level, in comparison to higher voltage grids. Asymmetric and nonlinear loads and generation units generate unbalanced distorted currents in the grids, which can significantly affect the voltage waveforms as well, due to the high equivalent grid impedance at this voltage level.

Considering asymmetries leads to more complex models of lines and cables and the necessity of a careful consideration of node grounding, which includes choosing the grounding model and how to parameterize it.

Simulating harmonics can drastically increase the model complexity, while simultaneously demanding a relatively small simulation time step. It is recommended to have a simulation sampling frequency of at least 20 times the maximum harmonic frequency that needs to be accurately represented [111]. At the same time, the simulation time step must be large enough to solve the complex differential equations of the model. This can result in a limited number of components that can be simulated in real-time.

3.1.4 A Summary of the Solutions for Real-time Simulation of Low Voltage Distribution Grids

In the previous subsections, the techniques and solutions that can help to tackle the challenges faced for real-time simulations of low voltage grids were discussed. A summary of these techniques and other methods proposed in literature is presented in Table 3.1. These methods can help to use optimally the computational capability of the real-time simulator, while preserving the model accuracy. For real-time simulation of medium- to large-scale low voltage distribution grids with a high penetration level of distributed resources, several of these techniques have to be combined in order to avoid overruns and achieve an acceptable model accuracy. Reducing the simulation time step of the grid model, can also help with the stability of a PHIL setup, as shown later in subsection 3.2.2.

Table 3.1: Possible techniques and solutions for real-time simulations of low voltage distribution grids with a high share of distributed energy resources.

Solution type	Solution	Conditions for applying
Low voltage grids model decoupling techniques	Reducing simulations time step to match the longest line	Long line (≥2 km) available
	Adding an artificial delay [103]	Accurate transient results not required
	Adding spurious shunt capacitance after the transformer [102]	Accurate transient results not required
	Using partition-based solvers such as SSN [109]	The solver is available
Low voltage grids model simplifications techniques	Neglecting lines' capacitances [117]	Short, lightly loaded lines and symmetric grid
	Neglecting lines' coupling inductances [106]	Symmetric grid
	Using single-phase models [118]	Symmetric grid
	Using phasor-based [119] or quasi-dynamic simulators [120]	Focus on slow grid transients
Converters real-time simulation techniques	Using average or phasor models [110]	Focus on slow grid transients
	Separating the controllers on separate core [121]	Relatively small simulation time step
	Decoupling the models at the DC-Link [122]	Large DC-link capacitor
	Using switching event compensation methods [111]	Switching frequency up to 10 kHz
	Co-simulation with the FPGA and the CPU [114]	Switching frequency ≥10 kHz

Due to the aforementioned challenges, low voltage grid models are often simplified to reduce the model execution time. In [118], authors have simulated the CIGRE low voltage benchmark [123] in real-time, only as a single-phase grid, assuming a fully symmetric operation of the grid. In addition, the distributed energy resources have been represented only by simple PQ sources. A 174-node distribution grid model was reduced to an equivalent 7-bus model in [20], in order to simulate the model in real-time with a simulation step of 50 μs. Another example is presented in [106], where the unbalanced coupled PI-lines are replaced with simple RL branches with shunt capacitors for model simplification of the IEEE 123 node feeder. In [101], a large distribution grid with 615 nodes has been simulated in real-time together with a microgrid using the state-space nodal solver, without any major simplifications. Among the available solutions, the use of state-space nodal solver can significantly extend the possibilities of real-time simulation of large low voltage grids, without model simplifications or compromising the model accuracy. Therefore, this method is applied later in Chapters 6 and 7 to simulate the studied low voltage grids and microgrids in real-time.

In addition to real-time simulations of power systems, real-time simulations of power system components, such as the energy storage systems can accelerate their design, analysis and validation [96], [100]. This is discussed later in Chapter 5, where a real-time simulation of a FESS for grid integration studies is provided.

3.2 Power Hardware-in-the-Loop (PHIL) Testing of Distributed Energy Resources

Real-time simulations enable a hardware to interact with the simulation models via I/O modules, which is considered a milestone in model-based design, and testing of new components and controllers [96]. In the 20th century, this was only possible with highly costly analog test benches to mimic the real system. Today, digital real-time simulators do the same task at much lower prices and with a higher accuracy and flexibility [107]. Depending on which part of a system is represented by the real hardware and which part by its real-time model, there are different concepts of using real-time simulations, as shown in Table 3.2.

Table 3.2: Different testing concepts using real-time simulations.

		Environment	
		Real-time Model	**Actual Hardware**
Controller or Component	**Real-time Model**	Software-in-the-Loop (SIL) testing	Rapid control prototyping
	Actual Hardware	Controller Hardware-in-the-Loop (CHIL) testing Power Hardware-in-the-Loop (PHIL) testing	Field testing

In Controller Hardware-in-the-Loop (CHIL) testing, only low-power signals are exchanged between the real-time simulator and the hardware under test, which is suitable for testing controllers and protection relays. Some examples include CHIL testing of energy management systems [124] and distributed voltage controllers [125]. In CHIL testing, the hardware under test receives signals as it would receive in the field and reacts accordingly. Power Hardware-in-the-Loop (PHIL) testing is an extension to the CHIL testing, in which significant electrical power is exchanged with the hardware in real time using power amplifiers. This is an approximate replication of field testing of power system components, at much lower expenses and risks and with a much higher flexibility and repeatability. Moreover, it allows to test the hardware in extreme conditions such as faults or frequency deviations, which is often not allowed or possible in field tests. PHIL testing can be a powerful tool to investigate the behavior of distributed energy resources and energy storage systems in various grid conditions. It has been shown that PHIL testing of distributed energy resources can show existing issues, that it is difficult to see using pure simulations [18]. This can be due to the difficulty in representing system nonlinearities in the model, such as delays, and component parameter variation. A schematic diagram of PHIL testing for a converter-interfaced distributed generation or an energy storage system is shown in Figure 3.2.

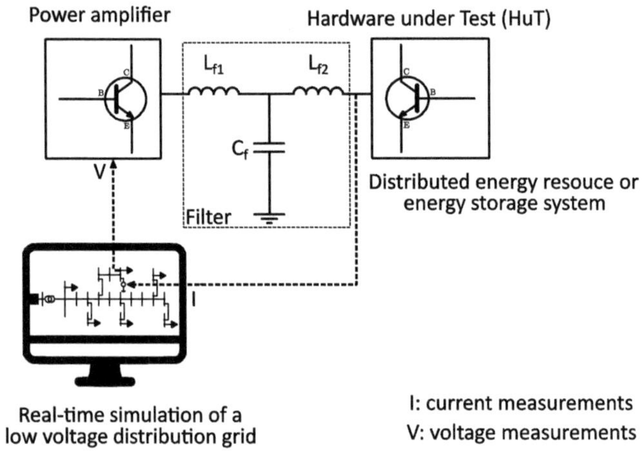

Figure 3.2: Schematic diagram of a typical PHIL setup for distributed energy resources.

There are several important steps that must be taken prior to any PHIL testing. These are listed in Table 3.3 and explained in the following subsections, which can serve as a guideline for carrying out PHIL tests. To summarize, firstly, the hardware under test and the grid under study should be modelled and simulated in real-time, and the models should be validated, when possible. The next step is the evaluation of the stability and accuracy of the PHIL setup, in order to avoid physical damage to the PHIL setup and the hardware under test. Finally, PHIL tests can be executed with adequate safety measures.

Table 3.3: A step-by-step guideline for PHIL testing for grid integration studies [118].

Steps	Tasks
Step 1: Test Case Development	1.1) Modelling and real-time simulation of the hardware under test.
	1.2) Modelling and real-time simulation of the grid under study.
	1.3) Verifying the real-time simulation models, when possible.
	1.4) Defining and real-time simulation of the test scenarios with the simulated model of the hardware under test.
Step 2: Stability and Accuracy Evaluation	2.1) Deriving open-loop and closed-loop transfer function of the power hardware-in-the-loop setup.
	2.2) Accuracy and stability evaluation of the power hardware-in-the-loop setup and choosing the interfacing method.
	2.3) Providing protection in case of hardware failure or instability.
Step 3: Power hardware-in-the-loop Testing Execution	3.1) Connecting the hardware under test using the chosen interfacing method for closing the loop.
	3.2) Running the real-time models and closing the loop at low power transfer.
	3.3) Simulate the selected scenarios and conduct power hardware-in-the-loop testing.

3.2.1 Test Case Development

Firstly, the hardware under test should be modelled and simulated to understand its behavior prior to testing the physical system. This helps to avoid unexpected behavior of the hardware during the tests and to design suitable and safe test scenarios. The level of the details required for this modelling depends on the intended type of study. Although offline simulation is also beneficial, real-time simulation of the hardware under test has the advantage that it can be directly inserted in the real-time simulation model of the grid, in which the actual hardware is later tested. The developed models can be validated with measurements from the real hardware, when possible.

Next, the grid, in which the hardware is planned to be tested, should be implemented in a real-time simulation environment. The details of the grid model, and the simulation step time should be suitable to represent a real power system for the intended study. As mentioned, it is recommended that the simulation sampling frequency should be at least 20 times greater than the maximum frequency of transients that needs to be represented with an acceptable accuracy, which is also known as model bandwidth [111]. Figure 3.3 shows the minimum simulation time step according to the model bandwidth. As seen, the model bandwidth also determines the minimum amplifier bandwidth, which helps selecting the appropriate power amplifier for the PHIL setup. Moreover, the maximum open loop delay for a stable setup is shown according to the model bandwidth. The stability of the PHIL setup is discussed in the following subsection.

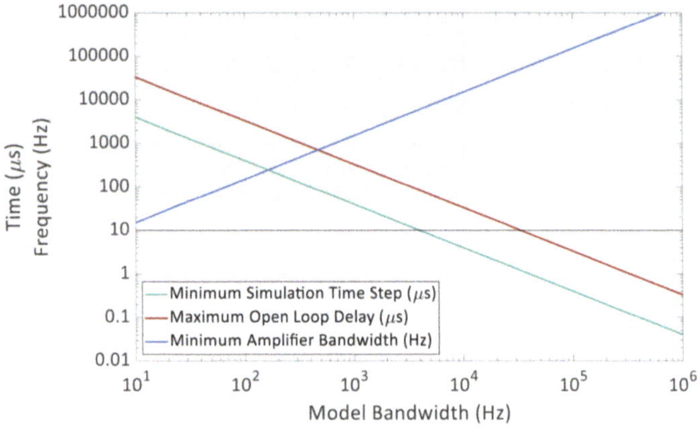

Figure 3.3: Real-time simulation and PHIL testing requirements according to the model bandwidth, defined as the maximum frequency of the harmonics or system dynamics to be represented accurately [126].

Lastly, the scenarios for the PHIL testing should be defined. These scenarios should be first simulated with the developed models of the hardware and the grid, preferably, in a real-time simulation environment. By doing so, one can have a good estimate of the behavior of the hardware under test in each test scenario. The delays and latencies associated with the PHIL setup can also be added to the simulation models to increase the simulation accuracy of the PHIL tests. The role of these delays and latencies on the stability and accuracy of the PHIL setup are discussed next.

3.2.2 Stability and Accuracy Evaluation

In order to replicate the direct connection of the grid and the hardware under test, an interfacing algorithm is required to exchange signals, as demonstrated in Figure 3.4. In one common approach, known as the ideal transformer method, the voltage values from the real-time simulation are amplified and fed to the hardware under test. The consequent current flowing through the hardware are measured and fed back to the simulation to close the loop. In Figure 3.4, Z_{eq} is the equivalent grid impedance in the real-time simulation environment, and Z_{HuT} is the impedance of the hardware under test.

The errors, time delays, and nonlinearities in different components of the PHIL setup can lead to system instability and inaccurate results [98]. For instance, the use of a power amplifier introduces additional control loops, errors, time delays, and dynamics in the system and limits the loop bandwidth [127]. Measurement systems can also cause delays due to filtering, A/D and D/A conversions, and signal transmission. A major factor contributing to the stability and accuracy of the setup is the simulation time step of the real-time simulator, where the methods introduced in Table 3.1 can be used to reduce simulation time step and increase the stability of the setup.

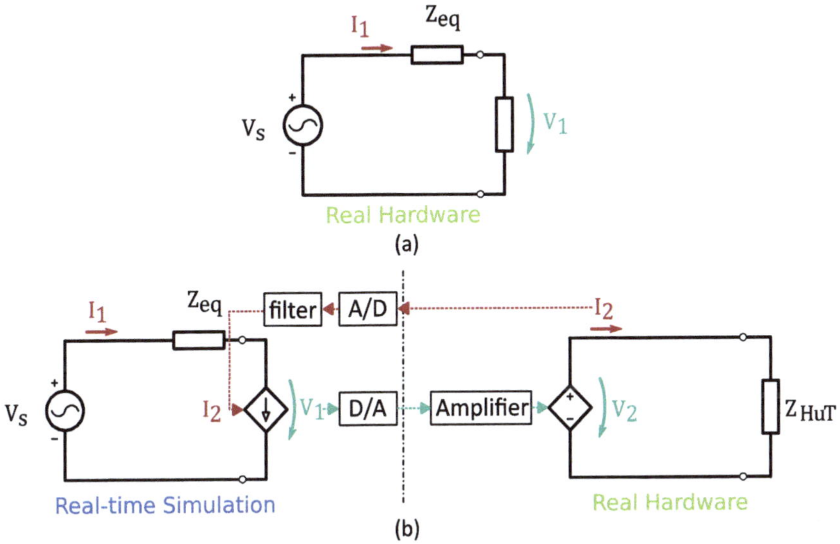

Figure 3.4: (a) Field test setup (ideal case). (b) PHIL setup using the ideal transformer method.

Therefore, the stability and accuracy of the PHIL setup with the hardware under test should be evaluated prior to conducting the experiments. Commonly, the stability of the system is analyzed using stability criteria from linear control theory, such as the Nyquist or Bode stability criterion, or using a transient simulation model of the setup. The accuracy of the system can also be evaluated based on the methods proposed in [128].

Researchers have proposed numerous interfacing algorithms for PHIL testing, which also includes methods that are too complex for implementation, with limited applicability. It is common practice to choose a simple interface algorithm that can guarantee stability with a reasonable accuracy. An ideal interfacing algorithm should be accurate with a low settling time and delay, stable in all testing conditions, with a high bandwidth and a simple implementation. In practice, the ideal transformer method and the damping impedance method are most commonly used [118], [129]. These two widely used methods are explained briefly in the following subsections.

3.2.2.1 The Ideal Transformer Method (ITM)

The Ideal Transformer Method (ITM) is the most commonly used interfacing algorithm for PHIL testing of linear and nonlinear components [18]. This is due to its simple implementation, high accuracy, and the fact that it requires only current measurements as feedback. In this method, the simulated voltage is amplified and fed to the hardware under test and the current flowing through it is measured and fed back to the simulation. A schematic diagram of the ideal transformer method is illustrated in Figure 3.4 (b). From Figure 3.4 (b), the block diagram of a PHIL setup when using the ideal transformer method can be obtained, which is shown in Figure 3.5.

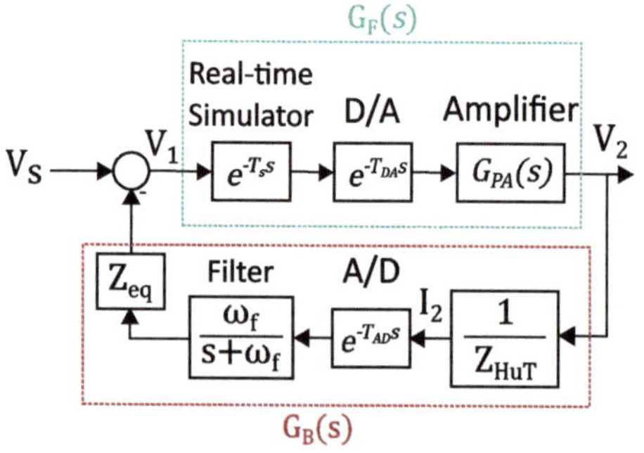

Figure 3.5: Block diagram of a PHIL setup when using the ideal transformer method.

In Figure 3.5, $G_F(s)$ and $G_B(s)$ are the feedforward and feedback transfer functions, respectively. These transfer functions can be easily calculated from the transfer function of each individual component of the PHIL setup. The feedforward transfer function $G_F(s)$ consists of the delay associated with real-time simulator, i.e., the simulation time step T_s, delays in signal transmission and D/A conversion, and the transfer function of the power amplifier $G_{PA}(s)$. The sum of all delays in the feedforward transfer function is represented by T_{d1}. Therefore, the feedforward transfer function $G_F(s)$, is calculated as

$$G_F(s) = G_{PA}(s)e^{-T_{d1}s}. \tag{3.1}$$

The feedback transfer function $G_B(s)$ on the other hand, also includes the ratio of the impedance of the equivalent grid impedance Z_{eq} to the impedance of the hardware under test Z_{HuT}. The delays of the current sensor and measurement, signal transmission, and D/A conversion should also be considered, which are all represented by T_{d2}. The feedback measurements can also include a first-order low-pass filter with a cut-off frequency of ω_f. Therefore, the transfer function of the feedback loop is calculated using

$$G_B(s) = \frac{\omega_f}{s+\omega_f}\frac{Z_{eq}}{Z_{HuT}}e^{-T_{d2}s}. \tag{3.2}$$

For analyzing the stability of the PHIL setup using the ideal transformer method, the open-loop transfer function $G_{OL}(s)$ can easily be calculated by multiplying $G_F(s)$ and $G_B(s)$, as shown in Eq. (3.3). The total loop delay T_d is defined as the summation of the feedforward and feedback delays, T_{d1} and T_{d2}. For simplification purposes, the power amplifier transfer function $G_{PA}(s)$ is also represented by a delay function, which its value is integrated into T_d. Therefore,

$$G_{OL}(s) = G_F(s)G_B(s) = \frac{\omega_f}{s+\omega_f}\frac{Z_{eq}}{Z_{HuT}}G_{PA}(s)e^{-T_{d1}s}e^{-T_{d2}s} = \frac{\omega_f}{s+\omega_f}\frac{Z_{eq}}{Z_{HuT}}e^{-T_d s}. \tag{3.3}$$

A system with the open-loop transfer function of Eq. (3.3) is a conditionally stable system, meaning it is stable only for specific gain values. This can be shown by the Nyquist diagram of the open-loop transfer function $G_{OL}(s)$, as illustrated in Figure 3.6. Using the Nyquist stability criteria has the advantage that the approximation of the time delay, using the Padé approximation [18], is not required. According to the Nyquist stability criteria, a system is unstable, if it encircles the point [-1,0], while the frequency increases on the Nyquist diagram [130]. As seen in Figure 3.6, with impedance ratio below one, the system becomes unstable. Therefore, the ideal transformer method can become unstable, when the ratio of impedance of the hardware under test to the equivalent grid impedance is less than one, i.e., $Z_{HuT}/Z_{eq} < 1$.

The parameters used for the stability analysis in this chapter are listed in Table 3.4.

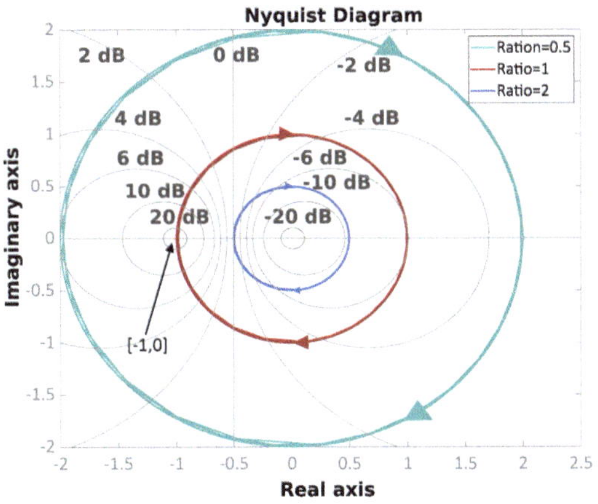

Figure 3.6: The influence of the system impedance ratio on the PHIL setup stability using the ideal transformer method.

Table 3.4: The parameters used for the PHIL stability analysis.

Parameter	Value
Simulation time step	10 µs
D/A and A/D conversion delay	3 µs
Power amplifier delay (for a linear amplifier)	3.1 µs
Sensors and measurement delay	2.8 µs
Equivalent grid resistance	4.5 mΩ
Equivalent grid inductance	14.4 µH
Hardware under test resistance	0.45 mΩ
Hardware under test inductance	1.44 µH

For low values of $Z_{\mathrm{HuT}}/Z_{\mathrm{eq}}$ the system can become stable by having a low-pass filter with an adequately low cut-off frequency of the filter [131], [132]. An example is illustrated in Figure 3.7 for a $Z_{\mathrm{HuT}}/Z_{\mathrm{eq}}$ of 0.1. As shown, a filter cut-off frequency lower than 2.5 kHz ensures system stability. However, being close to the point [-1, 0] indicates an oscillatory response [130], therefore further reduction of the cut-off frequency may be necessary. This can also be shown by calculating the gain and phase margin of the transfer function using the Bode stability criteria. As shown in Figure 3.8, a positive gain and phase margin is only achieved for a cut-off frequency of below 2.5 kHz. Similarly, since small phase and gain margins indicate an oscillatory closed-loop response, the cut-off frequency should be lower than 2.5 kHz.

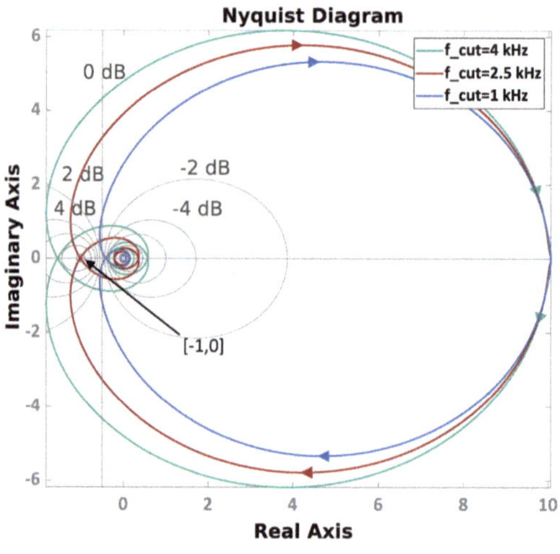

Figure 3.7: The influence of the filter cut-off frequency on the PHIL setup stability using the ideal transformer method for a $Z_{\mathrm{HuT}}/Z_{\mathrm{eq}}$ ratio of 0.1 using the Nyquist criteria.

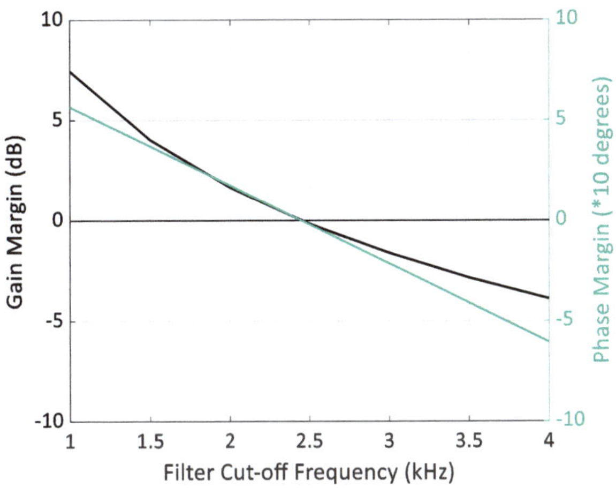

Figure 3.8: The influence of filter cut-off frequency on the power hardware-in-the-loop setup stability using the ideal transformer method for a $Z_{\mathrm{HuT}}/Z_{\mathrm{eq}}$ ratio of 0.1 using the Bode criteria.

However, using filters can lead to phase shifts and magnitude attenuations, which reduce the accuracy of the PHIL tests, and can lead to an unacceptable reactive power transfer [18]. Other methods to improve the stability of the ideal transformer method include adding additional components such as inductors [133], using the current-type ideal transformer method [129], interface compensation methods [113], and multi-rate interfacing [134]. Adding inductive components can improve the system stability but increases the hardware impedance, which obviously leads to a reduced accuracy and increased power losses, which is not preferred, in particular for in high-power applications.

The total loop delay T_d also has a significant effect on the system stability. The most dominant component of the loop delay is the simulation time step of the real-time simulation [135]. As discussed in section 3.1, the simulation time step is constrained by the model complexity and the required model bandwidth that needs to be simulated in real time. Consider the same impedance ratio of $Z_{\mathrm{HuT}}/Z_{\mathrm{eq}}$ ratio of 0.1 and a cut-off frequency of 1 kHz, which was shown to be a stable system for a simulation time step of 10 μs. Figure 3.9 depicts the phase and gain margins of the closed-loop transfer function as a function of the simulation time step, ranging from 0 to 100 μs. As seen, the system becomes unstable for simulation time steps of more than approximately 42 μs. Smaller simulation time steps lead to higher stability margins. Since using filters can reduce the model accuracy as previously discussed, reducing the simulation time step should be a priority for improving overall system stability.

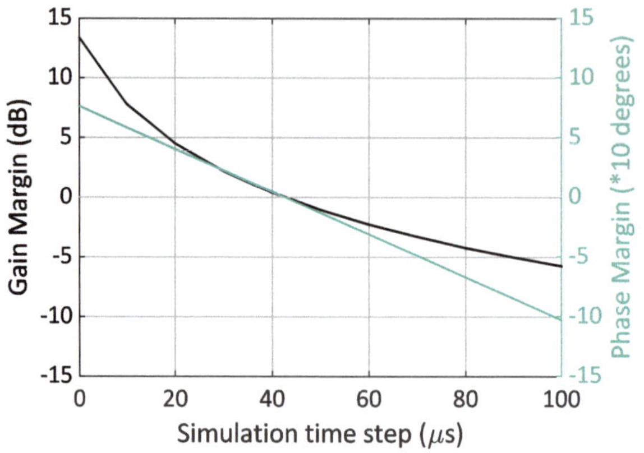

Figure 3.9: The influence of simulation time step on the power hardware-in-the-loop setup stability using the ideal transformer method for a $Z_{\mathrm{HuT}}/Z_{\mathrm{eq}}$ ratio of 0.1 using the Bode criteria.

3.2.2.2 Damping Impedance Method (DIM)

When the ideal transformer method cannot provide the adequate stability, other interfacing algorithms should be used. The Damping Impedance method (DIM) is another commonly used interfacing algorithm for PHIL testing, which can provide a higher stability margin in comparison to the ideal transformer method [129], [136]. A schematic of a PHIL setup using the damping impedance method is shown in Figure 3.10.

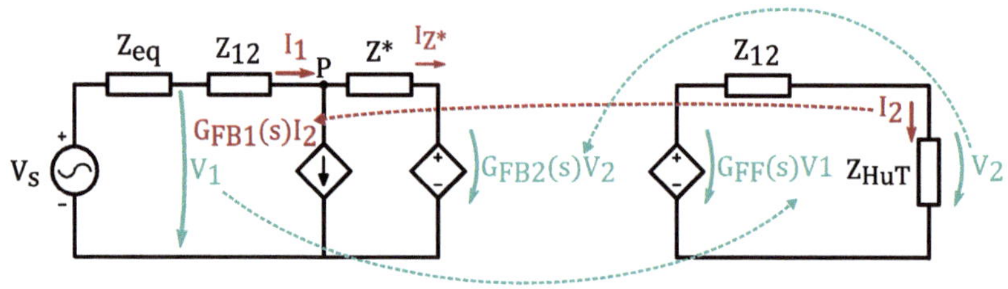

Figure 3.10: Schematic diagram of the Damping Impedance Method (DIM).

As shown in Figure 3.10, a linking impedance Z_{12} is added to both software and hardware side of the PHIL setup, while a damping impedance Z^* is added only to the real-time simulation. Figure 3.11 shows the block diagram of a PHIL setup using the damping impedance method. In Figure 3.10 and Figure 3.11, G_{FF} is the voltage feedforward transfer function, which includes the delay associated with signal transmission T_{d1} and the power amplifier $G_{PA}(s)$, while G_{FB1} and G_{FB2} are the transfer function of the current and voltage feedbacks, respectively.

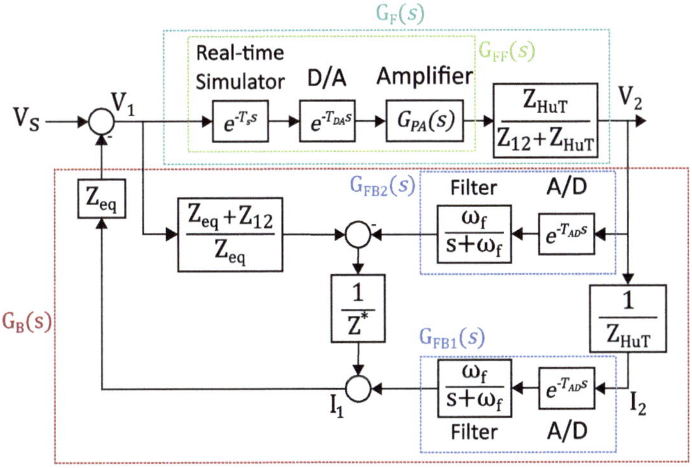

Figure 3.11: Block diagram of PHIL setup when using the damping impedance method.

The overall feedforward transfer function $G_F(s)$ using the damping impedance method can be easily calculated by considering the voltage divider shown on the right-hand side of the Figure 3.10. Therefore,

$$G_F(s) = \frac{V_2}{V_1} = \frac{Z_{HuT}}{Z_{12}+Z_{HuT}} G_{FF}(s) = \frac{Z_{HuT}}{Z_{12}+Z_{HuT}} G_{PA}(s)e^{-T_{d1}s}. \tag{3.4}$$

Calculating the overall feedback transfer function $G_B(s)$ is a bit more complex, as both the voltage and current measurements must be considered. As shown in [129], the overall feedback transfer function $G_B(s)$ is calculated using

$$G_B(s) = \frac{\omega_f}{s+\omega_f} \frac{\left(G_{FB1}\frac{Z^* Z_{eq}}{Z_{HuT}} - G_{FB2}Z_{eq}\right)}{(Z_{eq}+Z_{12}+Z^*)}. \tag{3.5}$$

By aggregating all system delays into one variable T_d and assuming that both voltage and current feedback filters are identical with the same low-pass filter, the open-loop transfer function G_{OL} using the damping impedance method is calculated by

$$G_{OL}(s) = G_F(s)G_B(s) = \frac{\omega_f}{s+\omega_f} \frac{Z_{eq}(Z^*-Z_{HuT})}{(Z_{12}+Z_{HuT})(Z_{eq}+Z_{12}+Z^*)} G_{PA}(s)e^{-T_d s}. \tag{3.6}$$

From Eq. (3.6), it is clear that when $Z^* = Z_{HuT}$, the magnitude of the open-loop transfer function becomes zero, which indicates a stable system, regardless of all the other variables such as the impedance ratio. However, this improved stability is achieved only when the damping impedance Z^* is equal to the equivalent impedance of the hardware under test Z_{HuT}. Therefore, the damping impedance method requires a good estimate of the impedance of hardware under test. This can be challenging in practice, in particular for active nonlinear components as hardware under test. In [129], [137], authors propose the use of online impedance identification techniques to match the simulated damping impedance with the impedance of the hardware under test. However, these methods are only applied to complex loads as the hardware under test and their application of power electronic-based devices is not yet investigated. They also add additional hardware and software complications to the PHIL setup. The damping impedance method is difficult to implement in comparison to the ideal transformer method and requires feedback from voltage measurements, which can introduce noise into the real-time simulation. Moreover, adding the linking impedance on the hardware side leads to power losses, which can be problematic for high-power applications. Therefore, this method should be used only if the ideal transformer method cannot provide a stable setup.

3.2.2.3 A Comparison of the Common Interfacing Algorithms

As discussed previously, there are several interfacing algorithms for PHIL testing proposed in literature, many of which are not discussed in the chapter. Each method can demonstrate improvements in terms of stability and accuracy in particular conditions, while often adding more complications to the setup. Table 3.5 provides a short comparison of the most commonly used interfacing algorithms.

Table 3.5: A comparison of the most commonly used interfacing algorithms for PHIL testing [98], [129], [138].

Interfacing algorithm	Advantages	Disadvantages
Ideal transformer method	Very easy implantation No extra hardware required High accuracy	Instable when $Z_{\mathrm{HuT}}/Z_{\mathrm{eq}} < 1$ High settling time when $Z_{\mathrm{HuT}} \approx Z_{\mathrm{eq}}$
Partial circuit duplication	High stability or accuracy	Either highly stable or highly accurate Accuracy depends on the linking impedance Power loss in the circuit Additional hardware parts needed
Damping impedance method	High stability when $Z_{\mathrm{HuT}} \approx Z_{\mathrm{eq}}$ High accuracy Low settling time	Impedance of the hardware is required Power loss in the circuit Additional hardware parts needed
Damping impedance method with impedance matching	High stability High accuracy Low settling time Hardware impedance not required	Increased complexity of the setup Increased model computation time Power loss in the circuit Additional hardware parts needed

3.2.3 Execution of the PHIL Testing

After developing the real-time simulation models and the test scenarios and a careful evaluation of the stability and accuracy of the PHIL setup, the hardware under test can be connected to the PHIL setup, using the chosen interfacing algorithm.

Even though a theoretical stability analysis can provide a valuable insight on the stability margin of a PHIL setup, care must be taken to avoid damage to the PHIL facility or the hardware under test, when unexpected instabilities occur [98], [113]. Errors in modelling and parameterization of the PHIL components can lead to an inaccurate stability analysis. For instance, it has been shown in [139] that the total loop delay can vary in time in a PHIL setup, as a result of the interaction between the discrete components of the setup. Therefore, hardware and software protective measures that can shut down the entire system, in case of instabilities or unacceptable high voltages and currents are mandatory.

For executing the PHIL tests, first the grid model should be running in real time. It is then recommended to close the loop with the measurement feedback when there is no or negligible power transfer between the hardware under test and power amplifiers. This can avoid hardware damages in case of incorrect settings or instability. When the loop is closed and the system shows a stable condition, the previously defined scenarios can be executed.

3.2.4 State-of-the-art in PHIL Testing of Energy Storage Systems

PHIL testing has been used to investigate the behavior of the different types of distributed energy resources under realistic conditions, including PV systems [98], [140]–[142], wind turbines [143], and energy storage systems. A summary of the literature review on the PHIL testing of energy storage systems is provided in Table 3.6.

In [20], PHIL testing of an inverter for a battery energy storage systems has been conducted in order to investigate the storage functionalities such as peak shaving and volt-watt control in a distribution grid. An adjustable DC source has been used in this work to emulate the battery. The applications of primary frequency control using a battery energy storage system has been tested in [19] using PHIL testing. It has been shown that the battery has a short response time of around 80 ms. A supercapacitor energy storage system has also been tested in a PHIL environment for frequency support in isolated areas [22]. Batteries and supercapacitors are combined to form a hybrid storage system and tested using PHIL testing in [23], where the supercapacitor is used to smoothen the high power gradients, whereas the battery provides power in the long term.

In case of PHIL testing of a FESS, limited experiments have been done with the focus on shipboard power systems for supporting pulsed loads in DC grids in [24], [25]. To our knowledge, PHIL testing of a FESS in low voltage distribution grids and testing grid functionality such as frequency support using PHIL testing has not yet been conducted.

Table 3.6: Literature review on PHIL testing of energy storage systems.

Type of ESS	Test Scenarios/Applications	Literature
Battery Energy Storage System	Peak shaving and volt-watt control.	[20]
	Primary frequency control	[19]
	Frequency regulation	[21]
	Interaction between an energy management system and the battery for active/reactive power compensation	[144]
	Supporting pulsed load in shipboard power systems	[145]
	Active and reactive control and operation monitoring	[146]
Supercapacitor Energy Storage System	Frequency support in a large, isolated power system	[22]
Hybrid Energy Storage System	Provision of primary frequency support and virtual inertia	[23], [147]
Flywheel Energy Storage System	Supporting pulsed load in DC shipboard power systems	[24], [25]

3.3 Description of the 1-MVA PHIL Infrastructure at KIT

As of December 2019, a 1-MVA PHIL testing facility has made operational by KIT-ITEP, as part of the KIT's EnergyLab 2.0.

An overview of this PHIL setup is illustrated in Figure 3.12. The core of this setup is an Opal-RT 5700 real-time digital simulator with 8 Intel Xeon processing cores and a Xilinx Virtex-7 FPGA on a VC707 board. In addition to the multiple analog and digital I/O boards on the simulator, the system also has 16 high-speed fiber-optic Small Form-factor Pluggable (SFP) sockets (up to 5GBps), which are used for connection to the amplifiers and to multiple OP4520 Kintex7 FPGA & I/O expansion units using Opal-RT's Multi-System Expansion (MUSE) link.

The software RT-LAB is used to compile the models in C programming language and load and execute the models on the real-time target. RT-LAB is also used for monitoring the real-time execution of the models, checking for overruns, assigning I/Os to model inputs and outputs, and changing model parameters, while the model is running on the target.

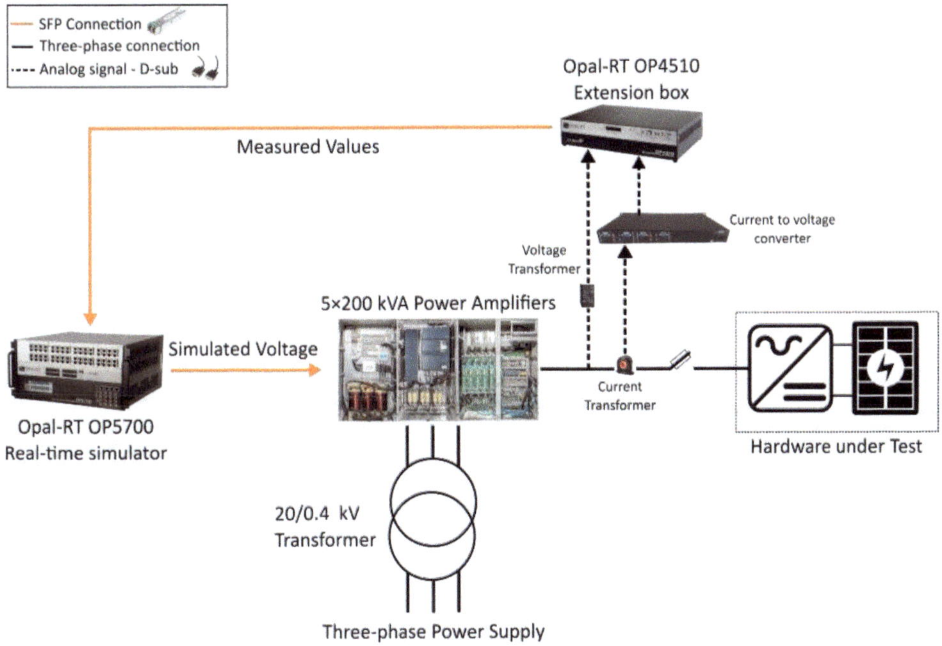

Figure 3.12: Overview of the 1 MVA PHIL setup at KIT's EnergyLab 2.0.

The voltage set points from the simulated grid are sent digitally to five 200 kVA switched mode amplifiers using the Xilinx's Aurora protocol. The use of this fast digital communication reduces the loop delay caused by signal transmission, by eliminating the D/A conversion unit and their corresponding anti-aliasing filters, and also the problems associated with noise and the signal grounding. The amplifiers are five 200 kVA COMPISO System Unit (CSU) GAMP6 from Egston Power. The nominal characteristics of each amplifier of the PHIL setup is presented in Table 3.7, and the inside view of each amplifier cabinet is shown in Figure 3.13. Each amplifier consists of 6 COMPISO Digital Amplifier (CDA), which can be connected and controlled in various configurations. To simulate a 4-wire three-phase power system, such as the low voltage grids in Europe, they are operated in 3-phase plus neutral mode, in which each phase composes of one CDA, and the neural has 3 CDAs for a high-power neutral connection.

The five Egston amplifier (CSUs), shown in Figure 3.14, can be operated in parallel or in series with each other to reach higher voltage and current ratings with a maximum power rating of 1 MVA. In this case, the set points are first sent an Egston control unit and from there to each individual amplifier.

Table 3.7: The characteristics of each Egston amplifier at KIT-ITEP.

Characteristics	Value
Nominal apparent power	200 kVA
Maximum AC voltage (phase-to-phase)	450 V_{RMS}
Amplifier bandwidth	5 kHz
Maximum AC current	252 A_{RMS}
Maximum DC voltage	725 V
Maximum DC current (unipolar operation)	900 A
Controller time step	4 μs

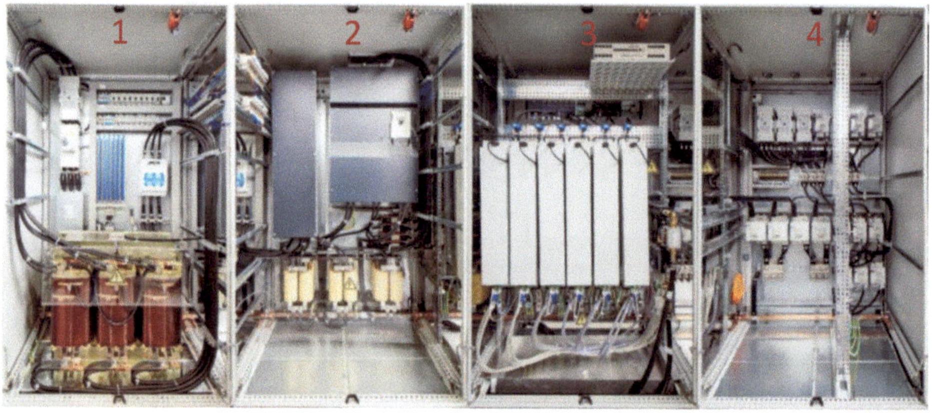

Figure 3.13: Inside of a 200 kVA COMPISO System Unit (CSU) power amplifier from Egston Power [126]. Cabinet 1 shows the grid connection and the isolation transformer, while the grid-side converter and its filter is in cabinet 2. The 6 COMPISO Digital Amplifier (CDA) are in cabinet 3, along with the Egston controller. Cabinet 4 shows the output contactors for the hardware under test.

Figure 3.14: Five 200 kVA GAMP6 Egston switched-mode power amplifiers (1 MVA in total).

For closing the loop for PHIL testing, the current and voltage measurements of the hardware under test are sent to two OP4520 Kintex7 I/O expansion units, which are equipped with an analog input card with 16 channels, 16 bits each, with 2.5 μs conversion time. The measurement values are also sent using high-speed fiber optics communication to the real-time simulator. Currents are measured using three categories of current transformers from the company Danisense, which can measure currents up to 640 A, 2 kA and 5 kA, respectively.

An emergency shutdown system has also been incorporated in the 1 MVA PHIL setup, which opens the amplifiers output breakers and de-energize the system, when necessary. In addition to the hardware protection schemes, software protection algorithms are also implemented on the real-time simulation to open the loop, send trip signals to the amplifier and hardware under test and force the set points to zero, when instable conditions or extremely high currents or voltages are observed.

3.4 Summary

This chapter introduced real-time simulations and its applications in power systems, including PHIL testing. For conducting PHIL tests, a real-time simulation of the grid under study is required. Therefore, a comprehensive overview of the challenges of real-time simulation of low voltage distribution grids, and the available techniques and solutions for tackling these challenges was provided. As discussed, methods that do not comprise the model accuracy such as using faster system solvers should be preferred, and often several methods should be combined to make the real-time simulation of complex grid models feasible.

Moreover, a guideline for conducing PHIL testing was presented. As discussed, having a model of the hardware under test and selecting an appropriate interfacing algorithm are necessary steps for designing safe and fruitful experiments. Among the available interfacing algorithms, the ideal transformer method should be preferred, whenever it can provide a stable setup. This is due to its simple implementation, high accuracy, and the fact that it does not require extra hardware components to be added, which can be a decisive factor for high-power PHIL setups. Moreover, it was shown that a reduction in the simulation step time of simulated grid can reduce the overall loop delay of the setup, and increase the stability margin.

Lastly, a literature review on PHIL testing of energy storage systems revealed that while such tests have been carried out for several storage technologies, a PHIL testing of a FESS in AC grids for applications such as frequency support has not yet been conducted.

4 Data-driven Allocation and Sizing of Energy Storage Systems

Among the major questions concerning the use of an ESS in distribution grids is how to determine its optimal location and characteristics, such as its capacity. It is desired to have the ESS in a location of the grid, where it can be the most beneficial to its stakeholders. After the location of the ESS is determined, it is also necessary to ensure that the installed ESS has the required characteristics to fulfil its main purpose of installation at that particular location. The main characteristics of an ESS include its rated capacity, rated power, and maximum ramp rate. Extensive literature has been published on the topic of optimal allocation and sizing of an ESS [148]–[150]. However, the proposed solutions can vary significantly depending on the storage technology, the intended application of the ESS, and the available grid and measurement data. Therefore, they are not always applicable to other use cases and applications. With the increasing amount of measurement data from low voltage distribution grids, data-driven techniques can provide new insights and solutions to the problem of allocation and sizing of an ESS.

In this chapter, two novel data-driven methods are introduced for the allocation and sizing of storage systems in low voltage distribution grids. Firstly, a new method for estimating the relative voltage sensitivity is introduced based on the concept of Mutual Information (MI) [26] from information theory. The method has been applied to find the most suitable location for a FESS for power smoothing applications in a large interconnected grid in southern Germany, with more than 1200 possible installation candidates. Next, a new method for sizing ESS based on historical measurement data is introduced, that uses reoccurring daily consumption patterns for the sizing, which are detected using the motif discovery algorithm [27]. The detected consumption patterns have been used for finding the characteristics of two types of storage technologies to form a hybrid ESS: A FESS for the application of power smoothing, and a high energy density ESS such as a Li-ion battery, for the application of peak shaving and load balancing. The sizing methodology has been applied for sizing these two types of storage technologies at four different low voltage distribution grids with different levels of PV penetration, and using data with different resolutions.

4.1 A Data-driven Method for Allocation of Flywheel Energy Storage Systems[1]

The optimal location for an ESS in a grid depends strongly on the main objective for its installation and the characteristics of the storage technology. An ESS can have various applications and functionalities in power systems. Therefore, there are numerous techniques proposed in literature for finding the optimal location of different types of ESS in power systems for different applications, including in [151]–[154]. Many of such methods are summarized and compared in [148]–[150]. However, the proposed solutions often focus on optimal allocation of battery energy storage systems for applications such as peak shaving, energy arbitrage, and voltage regulation, and they cannot necessarily be applied for other applications and storage technologies. The work focuses on allocating a high-speed FESS for the application of power smoothing in distribution grids.

A FESS has the ability to quickly and continuously smooth out short-term power fluctuations, as demonstrated in [155], [156], and [157], with no concern over its lifetime or capacity. Short-term power fluctuations in power systems can be caused by variation in loads, for example, industrial loads, or by intermittent generation from renewables, for instance by PV generation units. Power fluctuations at a relatively low amplitude are generally not a concern for the grid operator, but large power fluctuations could lead to significant voltage fluctuations in weak parts of the grid [158]. Fast voltage changes can be a power quality problem according to the European grid code DIN EN 50160 [159]. This is particularly the case in low voltage distribution grids with a low X/R ratio. As shown later in subsection 4.1.1, in such grids, the active power has as much as or even greater impact on the voltage, as the reactive power. In grids with a low X/R ratio, reactive power compensation can be inadequate for voltage regulation, unless an over-sized inverter is used [160], while active power compensation can significantly help to mitigate voltage deviations [161], [162]. A FESS can help smoothen the active power variations, which can potentially reduce the voltage fluctuations caused by the active power changes [163].

A FESS can be most useful at the grid connection points, where the voltage is more sensitive to active power changes. This can be quantified using the voltage sensitivity formulation, which is described in detail in the next subsection. The higher is the voltage sensitivity to active power variations, the higher is the impact of active power smoothing using a FESS on regulating voltage. The use of voltage sensitivity has been previously

[1] This section presents a revised version of the following publication: S. Karrari, M. Vollmer, G. De Carne, Klemens Böhm, Jörn Geisbüsch, "A Data-driven Approach for Estimating Relative Voltage Sensitivity," in the 2020 IEEE Power & Energy Society General Meeting, Montreal, 2020.

proposed for the optimal allocation of distributed energy resources to improve the system reliability indices [164], to increase the voltage stability margins [165], and to avoid over- and under-voltage incidents [166]. For the purpose of post-fault voltage recovery, a battery placement method based on voltage sensitivity has shown to be the most effective method in comparison to several other allocation strategies [167], [168]. Voltage sensitivity coefficients have also been used for designing advanced controllers for PV systems in order to improve the PV hosting capacity [169]. However, estimating voltage sensitivity in the aforementioned works requires a validated grid model, which is not always available, in particular for low voltage distribution grids.

This work proposes a new data-driven method for estimating the relative voltage sensitivity to active power changes, which is applied for finding the most suitable location for a high-speed FESS for power smoothing applications. The proposed method requires only measurement values at the points of interest, and does not require a grid model, which is a significant advantage. The suggested approach can also be used for other storage technologies and investigations, that require identification of the locations in a grid with a high voltage sensitivity to active or reactive power changes.

In the following subsections, after a short introduction to voltage sensitivity, the shortcomings of the classical method to calculate analytically the voltage sensitivity coefficients based on steady-state power flow equations are highlighted. Then, a novel data-driven approach has been proposed for estimating the relative voltage sensitivity. The proposed approach has been applied for finding the optimal location for a FESS among more than 1200 10/0.4 kV substations in a region in southern Germany, using the measurement data collected at the top candidates for the installation of the FESS.

4.1.1 Introduction to Voltage Sensitivity

Figure 4.1 shows a simplified power system with an equivalent impedance, consisting of an equivalent resistance (R_{eq}) and an equivalent reactance (X_{eq}) between the point of common coupling (Bus r) and the main supply (Bus s). All values are assumed to be in per unit.

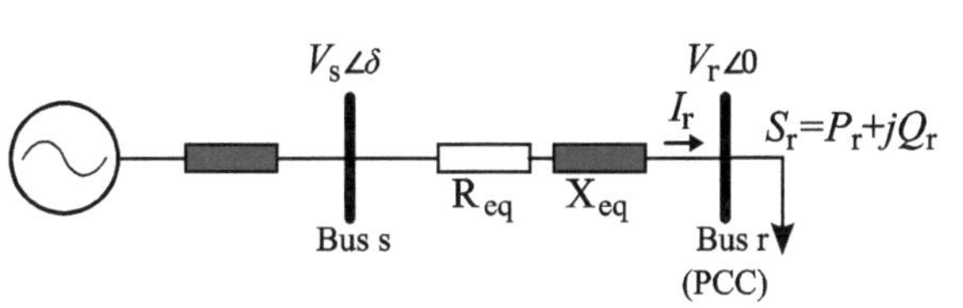

Figure 4.1: The single-line diagram of a simplified power system for calculating the voltage sensitivity.

The apparent power drawn from Bus r can be calculated as

$$S_r = P_r + jQ_r = V_r I_r^* = V_r \left(\frac{V_s e^{j\delta} - V_r}{R_{eq} + jX_{eq}}\right)^*. \tag{4.1}$$

By multiplying both the nominator and the denominator of the fraction on the right hand side of Eq. (4.1) by $(R_{eq} - jX_{eq})$, and separating the real and imaginary components, Eq. (4.1) can be divided into [170]

$$P_r = V_r \left(\frac{R_{eq}(V_s \cos(\delta) - V_r) + X_{eq} V_s \sin(\delta)}{R_{eq}^2 + X_{eq}^2}\right) \tag{4.2}$$

and

$$Q_r = V_r \left(\frac{X_{eq}(V_s \cos(\delta) - V_r) - R_{eq} V_s \sin(\delta)}{R_{eq}^2 + X_{eq}^2}\right). \tag{4.3}$$

It can be seen from Eq. (4.2) and (4.3) that the relationship between the voltage and the active and reactive power is nonlinear. Therefore, linearization around a single operating point is required to calculate the voltage sensitivity to active and reactive power changes. To calculate the voltage sensitivity to a small disturbance of ΔP_r and ΔQ_r around the operating point (P_{r0}, Q_{r0}, V_{r0}), Eq. (4.2) and (4.3) are linearized around this point. For doing so, P_r is replaced by $P_{r0} + \Delta P_r$, Q_r by $Q_{r0} + \Delta Q_r$, and V_r by $V_{r0} + \Delta V_r$, and δ is eliminated, with the trigonometric method described in [170]. The second-order terms such as ΔP_r^2 and ΔV_r^2 are neglected. Therefore, for instance, P_r^2 is replaced by $P_{r0}^2 + 2P_{r0}\Delta P_r$. After these mathematical manipulations, $\frac{\Delta V_r}{\Delta P_r}$ and $\frac{\Delta V_r}{\Delta Q_r}$ can be easily calculated, which are assumed to be the approximations of $\frac{\partial V_r}{\partial P_r}$ and $\frac{\partial V_r}{\partial Q_r}$ for small changes in the power around this operating point. Hence,

$$\frac{\partial V_r}{\partial P_r} \approx \frac{\Delta V_r}{\Delta P_r} = \frac{(R_{eq}^2 + X_{eq}^2)P_{r0} + R_{eq} V_{r0}^2}{V_{r0}\left(V_s^2 - 2V_{r0}^2 - 2(R_{eq}P_{r0} + X_{eq}Q_{r0})\right)} \tag{4.4}$$

and

$$\frac{\partial V_r}{\partial Q_r} \approx \frac{\Delta V_r}{\Delta Q_r} = \frac{(R_{eq}^2 + X_{eq}^2)P_{r0} + X_{eq} V_{r0}^2}{V_{r0}\left(V_s^2 - 2V_{r0}^2 - 2(R_{eq}P_{r0} + X_{eq}Q_{r0})\right)}. \tag{4.5}$$

The outputs of Eq. (4.4) and (4.5) are referred to as voltage sensitivity coefficients to active and reactive power changes, respectively, around the operating point (P_{r0}, Q_{r0}, V_{r0}). By assuming V_s and V_{r0} in these equations to be approximately 1 per unit, it results in

$$\frac{\partial V_{\mathrm{r}}}{\partial P_{\mathrm{r}}} \approx -\frac{\left(\mathrm{X}_{\mathrm{eq}}^2+\mathrm{R}_{\mathrm{eq}}^2\right)P_{\mathrm{r0}}+\mathrm{R}_{\mathrm{eq}}}{1+2\left(\mathrm{R}_{\mathrm{eq}}P_{\mathrm{r0}}+\mathrm{X}_{\mathrm{eq}}Q_{\mathrm{r0}}\right)} \tag{4.6}$$

and

$$\frac{\partial V_{\mathrm{r}}}{\partial Q_{\mathrm{r}}} \approx -\frac{\left(\mathrm{X}_{\mathrm{eq}}^2+\mathrm{R}_{\mathrm{eq}}^2\right)P_{\mathrm{r0}}+\mathrm{X}_{\mathrm{eq}}}{1+2\left(\mathrm{R}_{\mathrm{eq}}P_{\mathrm{r0}}+\mathrm{X}_{\mathrm{eq}}Q_{\mathrm{r0}}\right)}. \tag{4.7}$$

It can be seen from Eq. (4.6) and (4.7), that in case of $\frac{\mathrm{X}_{\mathrm{eq}}}{\mathrm{R}_{\mathrm{eq}}} \gg 1$, as in high voltage transmission grids, the voltage sensitivity to reactive power is much higher than the voltage sensitivity to active power, and therefore, the voltage variations depends mainly on the changes in the reactive power. However, for a low $\frac{\mathrm{X}_{\mathrm{eq}}}{\mathrm{R}_{\mathrm{eq}}}$ ratio, as common in low voltage distribution grids, the active power variations have also a significant impact on the voltage, since the voltage sensitivity to active power is also high. In fact, for the case of $\frac{\mathrm{X}_{\mathrm{eq}}}{\mathrm{R}_{\mathrm{eq}}} \approx 1$, the active power has as much influence on the voltage as does the reactive power, since Eq. (4.6) and (4.7) become identical. The difference in voltage sensitivity at different voltage levels is shown more intuitively in Figure 4.2, where the line gradients in Figure 4.2(a) and (b) determine the voltage sensitivity coefficients in high voltage transmission grids and low voltage distribution grids, respectively.

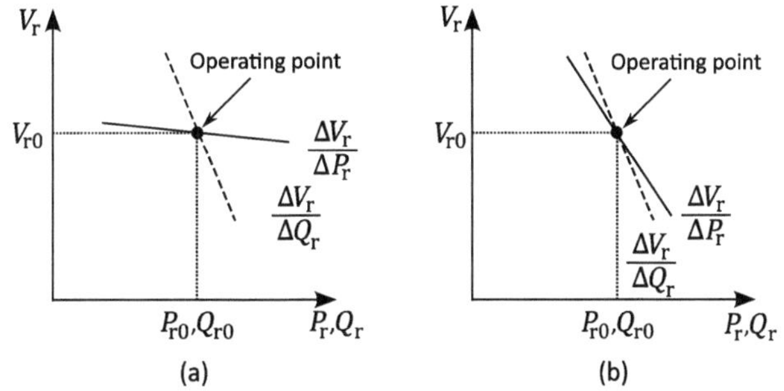

Figure 4.2: Different voltage sensitivity coefficients in (a) high voltage transmission grids, and (b) low voltage distribution grids (figures adopted from [170]).

For a large interconnected power system with numerous buses and lines, the voltage sensitivity coefficients are derived using one of the main three classical approaches described in [171], which all use the load flow calculations. The most common method, which uses the Newton-Raphson load flow formulation is described briefly in the following.

Assuming a balanced grid, by linearizing the load flow equations around one operating point we have

$$\underbrace{\begin{bmatrix} J_{P\delta} & J_{PV} \\ J_{Q\delta} & J_{QV} \end{bmatrix}}_{J} \begin{bmatrix} \Delta\delta \\ \Delta V \end{bmatrix} = \begin{bmatrix} \Delta P \\ \Delta Q \end{bmatrix}, \tag{4.8}$$

in which ΔP and ΔQ are small variations in the active and reactive power around the operating point, $\Delta\delta$ and ΔV are the resulted changes in the voltage phase and magnitude, respectively, and J is the Jacobian matrix of the grid. By multiplying both sides of Eq. (4.8) by the inverse of the Jacobian matrix, we have

$$\begin{bmatrix} \Delta\delta \\ \Delta V \end{bmatrix} = \underbrace{\begin{bmatrix} J_{P\delta} & J_{PV} \\ J_{Q\delta} & J_{QV} \end{bmatrix}^{-1}}_{S} \begin{bmatrix} \Delta P \\ \Delta Q \end{bmatrix}. \tag{4.9}$$

We define the inverse of the Jacobian matrix, S in Eq. (4.9), as the voltage sensitivity matrix of the grid. The elements of the voltage sensitivity matrix can be grouped into four submatrices as shown in Eq. (4.10).

$$S = \begin{bmatrix} S_{\delta P} & S_{VP} \\ S_{\delta Q} & S_{VQ} \end{bmatrix} = \begin{bmatrix} \frac{\partial\delta}{\partial P} & \frac{\partial V}{\partial P} \\ \frac{\partial\delta}{\partial Q} & \frac{\partial V}{\partial Q} \end{bmatrix} \approx \begin{bmatrix} \frac{\Delta\delta}{\Delta P} & \frac{\Delta V}{\Delta P} \\ \frac{\Delta\delta}{\Delta Q} & \frac{\Delta V}{\Delta Q} \end{bmatrix}. \tag{4.10}$$

In Eq. (4.10), S_{VP} corresponds to the submatrix of $\frac{\partial V_i}{\partial P_i}$, which is the partial derivative of the voltage amplitude at bus i with respect to the active power changes at the same bus, i.e., the voltage sensitivity to active power changes at the bus i, and

$$S_{VP} = \begin{bmatrix} \frac{\partial V_2}{\partial P_2} & \cdots & \frac{\partial V_2}{\partial P_n} \\ \vdots & \ddots & \vdots \\ \frac{\partial V_n}{\partial P_2} & \cdots & \frac{\partial V_n}{\partial P_n} \end{bmatrix}. \tag{4.11}$$

Similarly, S_{VQ} corresponds to the elements of $\frac{\partial V_i}{\partial Q_i}$ or the voltage sensitivity to the reactive power variations. The voltage phase angles and their sensitivities are generally not a concern at the low voltage level.

Up to this point, a balanced operation of the grid was assumed for the voltage sensitivity calculations. However, low voltage distribution grids often have asymmetric currents and voltages caused by single-phase loads and generation units. For an unbalanced three-phase system, the voltage sensitivity coefficients must be calculated for each phase individually with respect to the power flow of all three phases. Therefore, for bus i,

$$\frac{\partial V_\mathrm{i}}{\partial P_\mathrm{i}} = \begin{bmatrix} \frac{\partial V_\mathrm{ia}}{\partial P_\mathrm{ia}} & \frac{\partial V_\mathrm{i1}}{\partial P_\mathrm{i2}} & \frac{\partial V_\mathrm{ia}}{\partial P_\mathrm{ic}} \\ \frac{\partial V_\mathrm{ib}}{\partial P_\mathrm{ia}} & \frac{\partial V_\mathrm{ib}}{\partial P_\mathrm{ib}} & \frac{\partial V_\mathrm{ib}}{\partial P_\mathrm{ic}} \\ \frac{\partial V_\mathrm{ic}}{\partial P_\mathrm{ia}} & \frac{\partial V_\mathrm{ic}}{\partial P_\mathrm{ib}} & \frac{\partial V_\mathrm{ic}}{\partial P_\mathrm{ic}} \end{bmatrix}. \quad (4.12)$$

A more detailed analysis of voltage sensitivity calculations in unbalanced grids can be found in [171] and [172].

4.1.2 Limitations of the Classical Method for Calculating the Voltage Sensitivity

As shown, the classical methods for estimating the voltage sensitivity coefficients use the linearized load flow equations around a single operating point. This approach can provide accurate results only at the particular operating point, in which the calculations are conducted, and neglects the system nonlinearities. However, the grid's operating point varies significantly with time. It has been shown recently that the variations in the voltage sensitivity coefficients can be significant with the changes in the system operating point during a single day [172], [173]. Therefore, accurate estimation of the voltage sensitivity coefficients in every operating point is necessary, in particular when these values are used actively in controller actions, as in [169].

But most importantly, the classical method requires a validated and accurate grid model, or at least its Jacobian matrix, which is not always available. Grid models at the low voltage level are less common, in comparison to higher voltage levels. This can be due to the larger number of these grids in a single region, and the fact that they were considered traditionally the passive parts of the grid. Even if such models are available, they are often not validated nor continuously updated, and are often designed for the planning studies conducted by the grid operator for the expected load in the future. In addition, inverter-based distributed energy resources such as PV systems and their controllers have an impact on the aggregated load/voltage sensitivity [110], but are often not represented accurately in these load flow models.

Due to the aforementioned limitations of the classical approaches, data-driven methods for estimating the voltage sensitivity coefficients and without any knowledge of the grid model can be very advantageous, in particular for low voltage grids. There have been only two attempts for such solutions previously published in literature. Authors in [174], propose an algorithm for estimating the voltage sensitivity coefficients based on the smart meter data. Nonetheless, this method requires simultaneous measurements at different locations in the grid, and historical measurement data and load profiles to train the proposed algorithm. Another work proposes a near real-time estimation method to calculate these coefficients [172]. This work also uses a linear approximation during the measurement period for the estimation procedure. It should be noted that neither of the proposed approaches has been used for allocation studies.

In this thesis, a new data-driven approach has been proposed based on the concept of mutual information. Mutual information has the capability to capture and quantify the nonlinear dependencies between two variables, and it can also generate an output in real time [175]. It is suggested in this thesis that the mutual information can reflect the relative voltage sensitivity coefficients among different locations in a grid, and it only requires measurements at the points of interest. This can be used for applications such as allocation of distributed energy resources such as the allocations of the FESS for power smoothing applications, as demonstrated later in this chapter. Mutual information is described in the following subsection.

4.1.3 Introduction to Mutual Information (MI)

Mutual Information (MI), originated from the information theory [26], can capture and quantify the dependencies between two variables. It attempts to quantify "Given the values of one variable, how much does it help to guess the values of the other variable?". It reflects the expected amount of information that one variable can reveal about the value of another variable. In our use case, this means how much information about the variations of the active power drawn from a certain point in the grid, can be exposed by just looking at the voltage measurements. Mutual information is much more general than the Pearson correlation coefficient, since it can capture both linear and nonlinear dependencies [176]. Mutual information is zero, if and only if the variables are statistically independent. This is not true for the correlation coefficients.

Because of this ability to capture such complex dependencies, mutual information has been applied in different disciplines, such as machine learning [177] and medical imaging [178]. Mutual information has a great potential to be used in power systems, as highly nonlinear systems. However, to our knowledge, no application of mutual information in power systems has been reported, and this work is the first one to propose it.

Since mutual information is defined originally for probabilities, it requires some estimation to be used on measurement data, which are always limited in the number of data points. A well-known estimate, referred to as the KSG estimate after its creators, Kraskov, Stögbauer, and Grassberger [179], overcomes this challenge by using the distances between similar measurements, rather than estimating the distribution densities.

The KSG estimate is fast to compute and yields good results even for small and noisy measurement data [180].

Figure 4.3 presents an example of measurement points for defining the notations required for the KSG estimate of MI. Let $M = \{\mathrm{m}_1, \ldots, \mathrm{m}_\mathrm{n}\}$, in which $\mathrm{m}_i = (v_{\mathrm{a}i}, p_{\mathrm{a}i})$ be a collection of measurements of two variables, which are the voltage and the power in phase a in this use case. In Figure 4.3, M consists of 13 measurement points. For any $k \in \mathbb{N}^+$, the kth nearest neighbor of a measurement point m_i is represented by $kNN(\mathrm{m}_i)$, using the matrix distance of $dist(\mathrm{m}_i, \mathrm{m}_j) = \max(|v_{\mathrm{a}i} - v_{\mathrm{a}j}|, |p_{\mathrm{a}i} - p_{\mathrm{a}j}|)$. For instance, in Figure 4.3, the measurement point m_3 is the fourth nearest neighbor of m_5, in other words, $\mathrm{m}_3 = 4NN(\mathrm{m}_5)$. Furthermore, $c_{V_\mathrm{a}}(\mathrm{m}_i)$ and $c_{P_\mathrm{a}}(\mathrm{m}_i)$, referred to as the marginal counts, are the number of measurement points of voltage and power, respectively, that are not further from m_i than the $kNN(\mathrm{m}_i)$. For example, in Figure 4.3, $c_{P_\mathrm{a}}(\mathrm{m}_5) = 3$, i.e., there are three measurement point of P_a that their distance is not larger than the distance of $4NN(\mathrm{m}_5)$ or m_3. In a similar way, it can be seen that $c_{V_\mathrm{a}}(\mathrm{m}_5) = 7$.

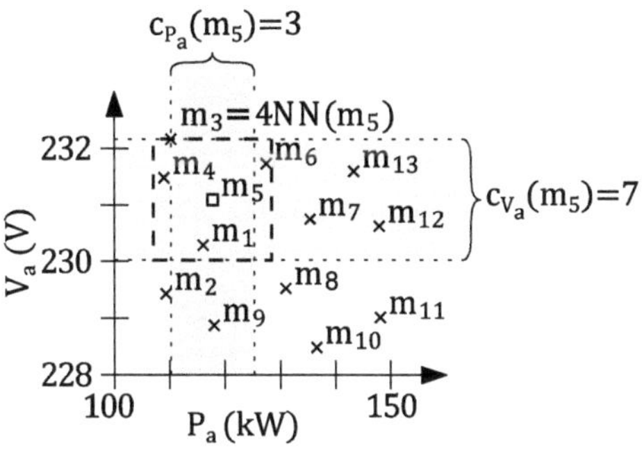

Figure 4.3: Examples of notations used for estimating the mutual information using the KSG estimate.

With these definitions, the mutual information using the KSG estimate for these set of measurements is calculated as [179]

$$MI(V_\mathrm{a}, P_\mathrm{a}) = \lambda(n) + \lambda(k) - \frac{1}{k} - \frac{1}{k}\sum_{i=1}^{n}\left(\lambda\left(c_{V_\mathrm{a}}(m_i)\right) + \lambda\left(c_{P_\mathrm{a}}(m_i)\right)\right), \tag{4.13}$$

in which,

$$\lambda(x) = \left(\sum_{m=1}^{x-1}\frac{1}{m}\right) - \gamma. \tag{4.14}$$

In Eq. (4.13) and (4.14), k is chosen to be 4, as recommended in literature [180], and γ equals to -0.577, which is the Euler-Mascheroni constant.

The value calculated using Eq. (4.13) is referred to as the mutual information score. It is proposed in this work that the mutual information score between voltage and active power measurements can reflect the relative voltage sensitivity to active power changes at different locations in a grid. Similarly, the mutual information score between the voltage and the reactive power can show the relative voltage sensitivity to reactive power variations. This is demonstrated in the next subsections, using real measurement data and the use case of allocation of a FESS in a low voltage distribution grid.

4.1.4 The Proposed Allocation Method for a FESS

In this section, a real case of allocation of a FESS for power smoothing applications in a grid in the southern Germany is presented, as a use case of the proposed method for estimating the relative voltage sensitivity using the mutual information scores.

The distribution system operator intends to install a high-speed FESS in its grid for improving the power quality. The system operator has limited the installation locations to the low voltage side of all the 10/0.4 kV substations, where there is sufficient utility-owned land and electrical and communication infrastructure for the FESS. This results in a total search space of over 1200 substations, as possible candidates for the installation of the FESS. The main application of the FESS is considered to be smoothing out the power drawn from the medium voltage grid in order to reduce the voltage deviations. Therefore, a location with a high voltage sensitivity to active power changes is desired.

To find the substation with the highest voltage sensitivity, a new allocation algorithm based on the concept of mutual information for estimating the relative voltage sensitivity is proposed. The proposed allocation algorithm consists of two main steps, which are described in detail in the following, along with the data collection procedure. The flow chart of the proposed allocation algorithm is shown in Figure 4.4.

4.1.4.1 Step 1: Selecting the Top Candidates for Measurements

Medium voltage grids are often monitored only at their interconnection with the high voltage grids, while low voltage grids are usually not monitored at all [181]. This is commonly justified by high expenses or being unnecessary, since they are large infrastructures, and were assumed to be the passive parts of the grid. Many distribution grid operators, collect measurement data only, if power quality issues or interruptions are reported. Despite the ongoing effort to improve the observability at these voltage levels in order to deal with the increasing share of distributed energy resources, the number of fixed measurement devices at these voltage levels is still low. This has been also the case for the medium voltage and low voltage grids investigated in this work. Therefore, first, measurement data have to be collected.

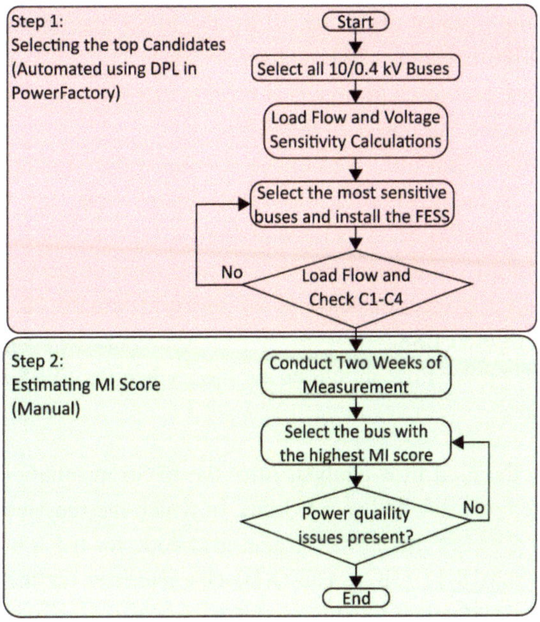

Figure 4.4: The flow chart of the proposed allocation algorithm for the FESS based on mutual information.

Since collecting measurement data from all 10/0.4 kV substations is not feasible nor practical, the substations with the highest potential for the FESS installation are selected for conducting measurements. As measurement data are not available at this point, the voltage sensitivity coefficients are calculated using the classical method based on the Newton-Raphson load flow calculations at the nominal operating point, as described in subsection 4.1.1. A static load flow model of the medium voltage grid in DIgSILENT PowerFactory is provided by the system operator, which is used for these calculations. It should be noted that this model is used only for planning studies of the grid operator.

This first step of the allocation algorithm is programmed using the DIgSILENT Programming Language (DPL) in DIgSILENT PowerFactory. Firstly, all the 10/0.4 kV substations are selected and the voltage sensitivity coefficients are calculated at the nominal operation point at all buses, using the method explained in subsection 4.1.1. The top nine buses with the highest voltage sensitivity to active power according to the model have been chosen as the most attractive candidates for conducting measurements and possibly the installation of the FESS. Next, the FESS is installed at each of these buses in the model, and the load flow calculations are repeated to check the following technical constraints (C1-C4):

- **C1:** After the installation of the FESS, the FESS should not exceed the grid's operational limits (ratings of transformers, lines, etc.) at its rated power, during both charging and discharging. This means, that the grid should be able to take up or provide the full power of the FESS.
- **C2:** The voltage limit of ±3 % of the nominal value should not be exceeded at any point of the grid after the FESS installation, for both charging and discharging of the FESS at its rated power.
- **C3:** Assuming 1.2 p.u. for the short-circuit current of the FESS, no interference with the latest protection settings of the grid should be observed after the installation of the FESS.
- **C4:** Reverse power flow should be avoided, when the FESS is discharging at its rated power.

In order to carry out the load flow analysis after the FESS installation, the bus at which the FESS is connected, is modelled as a PQ bus, in which the reactive power follows the Q(U) characteristics given in the latest German grid code for the low voltage grids, i.e., the VDE-AR-N 4105:2018-11 [182]. The Q(U) characteristic for the ESS connected to the low voltage grids is illustrated in Figure 4.5(b), together with the P(f) characteristic. In this figure, u_g and f_g are the voltage and frequency of the grid, respectively, Q_{max} is the maximum reactive power of the ESS, and P_{Emax} is its maximum active power. Different load flow models for an ESS can be found in [183].

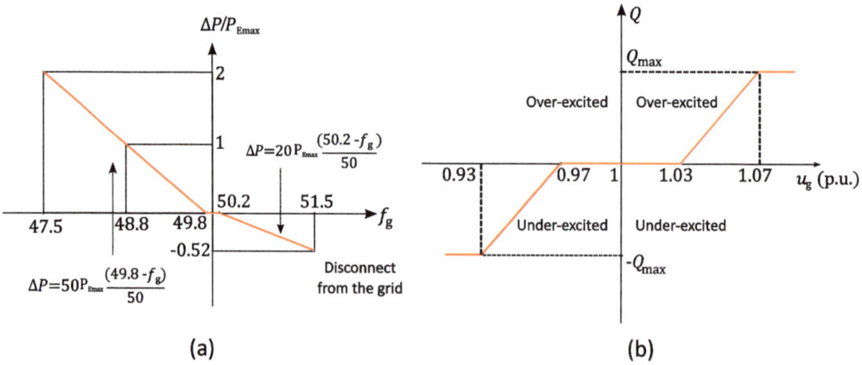

Figure 4.5: Active and reactive power control characteristics according to the German grid code VDE-AR-N 4105:2018-11. (a) The P(f) characteristic for frequency deviations, and (b) the Q(U) characteristic for voltage deviations.

If one or more of these constraints are violated after the installation of the FESS at a particular bus, the bus is replaced with the next bus with the highest voltage sensitivity to active power changes, and the constraints are checked again. There are also some non-technical constraints, such as having adequate space and infrastructure for installing the FESS. If the selected location does not have any of these conditions, the location is also replaced with the next most sensitive bus. The top nine candidates which fulfil both the technical and non-technical requirements are chosen for field measurements.

4.1.4.2 Measurement Setup at the Low Voltage Distribution Networks

In this subsection, the measurement setup used at the most attractive candidates for the FESS installation in a grid in southern Germany are described.

The measurements were conducted in a close cooperation with the distribution system operator of the region. An exemplary schematic of the measurement setup is illustrated in Figure 4.6. For each measurement, the measurement device was installed at the secondary side of the 10/0.4 kV transformer, in order to measure the power profile of the whole low voltage distribution grid, including all its feeders. Voltages were measured directly, while currents were measured using 800/5 A current transducers. The neutral current was measured separately.

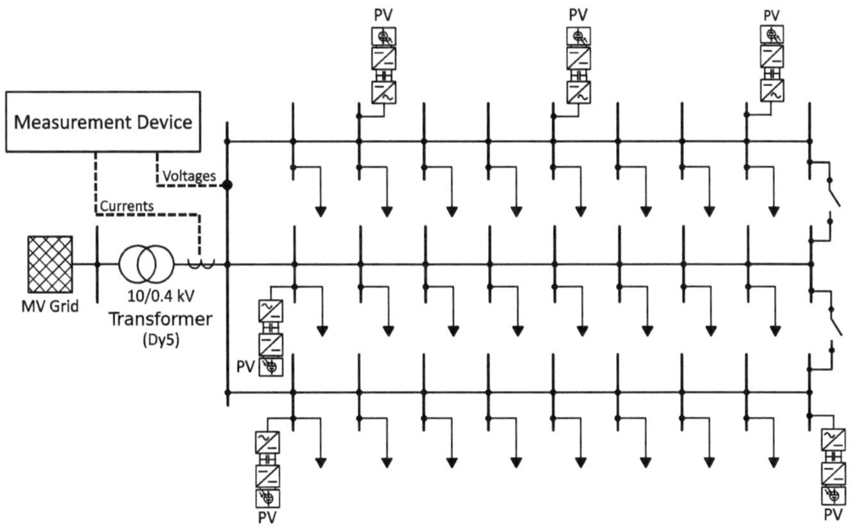

Figure 4.6: Schematic example of a measurement setup at the low voltage distribution networks.

The measurements were conducted using A. Eberle's power quality and disturbance recorder, "PQI-DA smart" [184]. This device samples the data at the high resolution of

41 kHz, but records the data on longer intervals, starting from every second. The high-resolution data is only recorded, if one of the predefined thresholds, determined by the power quality indices of the European standard DIN EN 50160 [159], are violated. When a continuous recording at a high resolution was required, the in-house built Electrical Data Recorder (EDR) [185] was used, which is capable of measuring and recording continuously at 25 kHz of sampling rate.

The measurement data were collected at the nine substations, which were chosen according to the step one of the allocation algorithm. A total of 223 days and 124 GB of data has been recorded. The measurements were conducted on low-voltage grids with different types of loads, including residential, commercial, industrial, and a combination of these types. Therefore, different patterns of load profiles were observed. Also, the low voltage grids have shown different penetration levels of PV generation, from no PV installations at all, to the case with enough PV generation to surpass the local demand at peak of generation. A minimum of two weeks of measurements have been conducted at the most attractive locations for the FESS installation.

From the perspective of power quality, no major violations of the European standard DIN EN 50160 [159] has been observed during the whole measurements. However, a series of interesting events have been recorded, which are presented in Appendix A.

4.1.4.3 Step 2: Estimating the Mutual Information Scores

For each bus, the mutual information scores between the voltage and active and reactive power measurements are calculated according to Eq. (4.13) using a window length of two weeks. The average of the mutual information scores of the three phases at the top nine candidates is shown in Figure 4.7.

As seen in Figure 4.7, bus 4, followed by bus 3 and 1, shows the highest mutual information scores between the voltage and the active and reactive powers, indicating a higher dependency of the voltage and the power at these points of the grid. This shows a higher voltage sensitivity to active and reactive power changes, in comparison to the other buses. Therefore, they can be more suitable for the installation of the FESS.

Bus 5 and 2 have the lowest mutual information scores, indicating a low voltage sensitivity to the active and reactive power changes. This can imply a strong grid connection, where active and reactive power variations have less impact on the voltage, when compared to other candidates. In addition, the results for bus 5, 7, and 8 show a higher mutual information score between voltage and active power, in comparison to the voltage and reactive power. This reflects a higher voltage sensitivity to active power in comparison to the reactive power, which is the characteristic of a low X/R ratio.

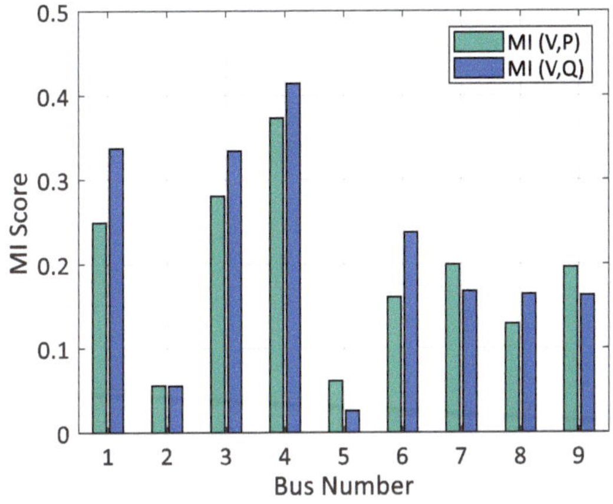

Figure 4.7: Comparison of the mutual information (MI) scores of the top nine candidates for the FESS installation.

4.1.5 Evaluation of the Proposed Allocation Method

In order to evaluate that the mutual information scores can reflect the relative voltage sensitivity at the different locations in a grid, the voltage and active power measurements, collected at bus 4 and 5, i.e., the buses with the highest and lowest mutual information scores between voltage and active power, are illustrated in Figure 4.8. Measurement data are shown for a period of two days with a 1-minute data resolution. The measurements are shown for bus 4 and 5 specifically, as their differences can be seen more clearly, compared to two buses with similar mutual information scores.

By observing Figures 4.8(a) and (b), it can be seen that the active power changes have a clear and significant impact on the voltage at bus 4, and these two variables greatly depend on each other. It can be observed, that the voltage variations follow the changes in the active power in the opposite direction, and their general pattern seems very similar. This indicates a weak connection point of the grid. For bus 5 with the lowest mutual information score, it can be observed from Figures 4.8(c) and (d), that the effect of the active power variations on the voltage is not easily recognizable. The dependency between the voltage and active power seems to be much lower, when compared to bus 4, and the patterns in the active power changes cannot be easily observed in the voltage. This is the characteristic of a stronger connection point of the grid, in comparison to bus 4. Similar results can be observed by comparing other buses to each other and by comparing the sensitivity coefficients of reactive power. Therefore, it can be concluded the proposed data-driven method based on the concept of mutual information can reflect the relative voltage sensitivity to active and reactive power changes at different locations in the grid.

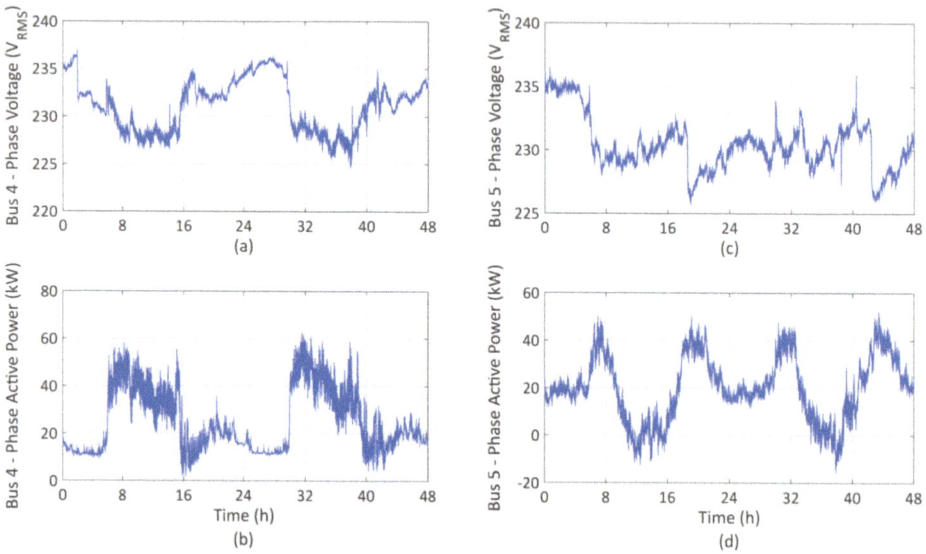

Figure 4.8: Measurement data at bus 4 and 5 illustrated for a period of two days with a 1-minute data resolution. (a) Bus 4 voltage, (b) bus 4 active power, (c) bus 5 voltage, and (d) bus 5 active power. Bus 4 has the highest mutual information score between voltage and active power, while bus 5 has the lowest score.

The proposed approach can be used for a variety of applications such as the allocation of distributed energy resources and ESS, in particular for improving the voltage quality. For the presented use case of allocation of a FESS in a grid in southern Germany, the proposed allocation algorithm has shown that the bus 4, followed by bus 3 and 1, has the highest potential for installing the FESS, as a higher relative voltage sensitivity to active power variations is observed (see Figure 4.8). However, just having a high voltage sensitivity to active power changes is not adequate in order to effectively benefit from a FESS. It is also important to have short-term active power fluctuations with high peaks, such as the ones caused by pulsed power loads, at the selected location in the grid. Such peaks in locations with a high voltage sensitivity to active changes can lead to significant fast voltage changes, which is a power quality violation. Where there is a high voltage sensitivity to active power variations, but no sharp active power fluctuations are present, significant fast voltage changes caused by active power changes will not be observed.

Therefore, the measurement results at the top three candidates with the highest mutual information scores, bus 4, 3 and 1, have been compared in terms of having short-term active power variations. Measurements result show, that the active power at bus 1 shows the highest number of short-term active power peaks. Due to the high voltage sensitivity at this location, the highest number of voltage deviations above 3 % is also observed at this location during the whole measurement period. A total number of 76 incidents of voltage deviations above ±3% of the nominal value were recorded during the measurement period of two weeks, which were caused by sharp active power changes. There are fewer voltage deviations recorded at bus 3 and 4, and none of them were caused by active power changes. An example of variations of the voltages and currents during several hours of measurements at bus 1 is shown in Figure 4.9. It is clear from Figure 4.9 that a pulsed power load is connected at this point in the grid. Since the voltage sensitivity to active power changes is also relatively high at this bus, these pulsed power loads cause significant fast voltage changes, as seen in Figure 4.9.

In conclusion, bus 1 is chosen as the most attractive location for the FESS installation among more than 1200 substations in the studied German grid. The grid can benefit the most from this storage technology at this particular location, as it is a weak point in the grid, and a large pulsed power load is also connected to it. The FESS can smoothen the power at this substation, by providing the short-term power requirements. Subsequently, the voltage deviations caused by the power pulsed load can be avoided.

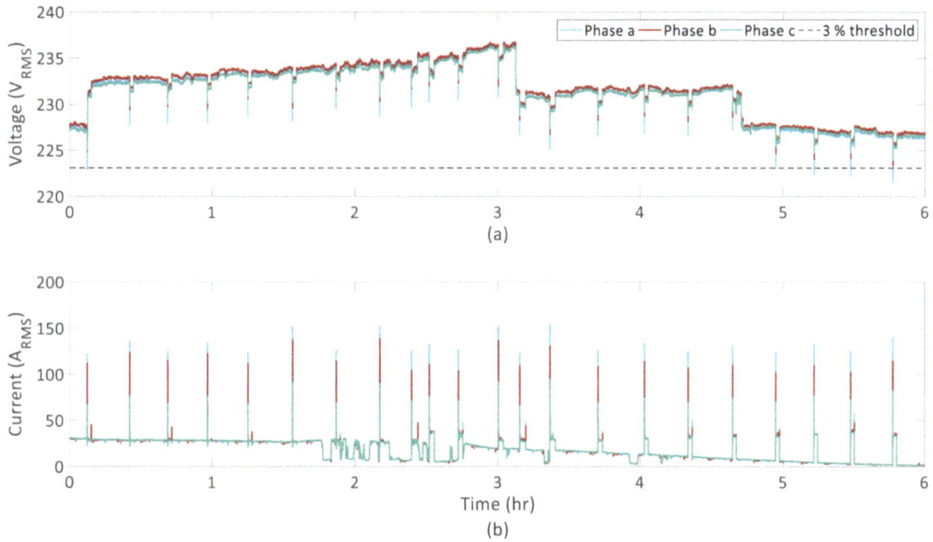

Figure 4.9: An example of the variations in the voltages and currents at bus 1, the most attractive candidate for the FESS installation during a period of 6 hours.

4.2 A Data-driven Method for Sizing ESS using Standard Patterns[2]

A high penetration level of PV generation in low voltage distribution grids can cause several challenges for the distribution system operator. The peak in PV generation appears around noon, when the demand is often low in residential areas with a high full-time employment rate [186]. This can lead to a reverse power flow from the PV installations towards the medium voltage grid during noon, which can potentially lead to over-voltage issues in the low voltage grids, due to the low X/R ratios [187], [188]. This can limit the PV hosting capacity of the low voltage grid for new PV installations [169], [187]. An example of a daily power profile of a low voltage distribution grid with a high share of PV generation is shown in Figure 4.10. This data is collected at a 10/0.4 kV substation in southern Germany during a summer day in 2018, using the setup described in subsection 4.1.4.2. It can be seen, that the active power drawn from the substation becomes negative after around 11:00 until approximately 15:00. The problem of reverse power flow and non-coinciding demand and generation can be solved by installing an ESS with high energy density, such as a lithium-ion battery energy storage system, if sized properly.

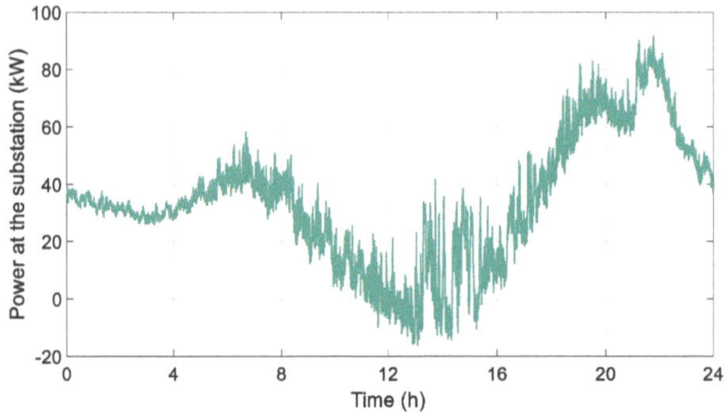

Figure 4.10: Example of a daily power profile of a low voltage distribution grid with a high share of PV generation during a summer day of 2018 in southern Germany.

In addition, as discussed in the previous section, the variations in the active power can significantly impact the voltage in grids with a low X/R ratio. Therefore, the fast variations in the solar irradiance, e.g., caused by passing clouds, which cause the power

[2] This section presents an extended verion of the following publication: S. Karrari, N. Ludwig, M. Noe, V. Hagenmeyer, "Sizing Centralized Energy Storage Systems in Distribution Networks Using Standard Patterns," in IEEE PowerTech 2019, Milan, 2019.

output of the PV systems to fluctuate, or variable industrial loads, can cause rapid voltage deviations in such grids. These rapid voltage fluctuations might violate the power quality threshold for the maximum permissible fast voltage changes [189], such as the 0.05 p.u. limit, given in the European grid code DIN EN 50160 [159]. Examples of large active power fluctuations can be observed in the measured power profile of Figure 4.10, between 13:00 and 15:00. This fast active power changes can be compensated using a high power density ESS with a large number of cycles such as a FESS, if it has the sufficient rated power and capacity.

The objective of this section is to derive the energy storage characteristics from the recorded data. As seen in the related work, the first step for sizing an ESS based on historical measurement data is to select a suitable data set as the input for the sizing methodology. The data resolution and the chosen period of the recorded data can both impact the sizing outcome. While 15-minute [190] or even hourly data [191] are frequently used in storage sizing studies, it is recommended to use higher resolution data, such as a data resolution of 1 second, in order to study the use of an ESS in the presence of intermittent PV generation in low voltage grids [158]. Similarly, another study concluded that the use of the full frequency spectrum for storage sizing studies is highly recommended [192]. However, using high-resolution data, or even low-resolution data on a long horizon (several years) leads to large data sets. This can drastically increase the computation time required for solving the sizing problem, in particular for complex nonlinear optimization frameworks, which can make the problem intractable [193]–[195]. In order to avoid dealing with such large data sets, it is common practice to choose some specific short period, e.g., one day, as a representative of the measurement data for the sizing study, or to reduce the data resolution by aggregating the data. However, the choice of a representative day is often not clearly explained, nor properly justified. Frequently, an arbitrary day or some worst-case scenario, e.g., a day with high PV generation and low demand, or a day with low PV generation and high load, is selected. These approaches can potentially lead to over- or under-sized ESS. For instance, in [196], a random day is chosen as the load profile for the sizing study, which can lead to very specific storage requirements that are not suitable for other days. Due to the lack of high-resolution data, 1-minute data are artificially generated from hourly data in [197], but again, the data of a typical day is used, and it is not clear how this typical day is selected. A similar approach has been reported in [198], where the logic behind choosing the typical day is not provided by the authors. In [192], a random selection of days is used to calculate the storage ratings. The alternative solution, which is reducing the data resolution is also not recommended, as discussed earlier. As an example, authors in [199] formulate the sizing problem as a convex optimization problem, but convert the 10-second load data to 15-minute data to ease computation over a year, which eliminates the high-frequency components of the power measurements.

To properly address this issue in sizing studies, clustering techniques such as K-means clustering [194], [200], [201] and fuzzy C-means clustering [202] were used to group similar daily power profiles and generate representative ones, that can be used for the sizing problem, instead of the whole data set. This is a significant improvement to the arbitrary choice of a typical day. However, when using clustering techniques, it is required to specify beforehand, how many clusters are needed to be found. Depending on the number of clusters and selected features, this can result in power profiles being grouped into one cluster, while not being necessarily similar, since often simple features such as the peak or mean value is used for the clustering (see [202] as an example). These approaches can also generate clusters of power profiles that occur very seldom, which cannot represent the data set properly. It should also be noted that the related works which use clustering techniques all use hourly data for long term scheduling problems and high-resolution data is not used. Table 4.1 provides a summary of the methods used in literature for reducing the data size for sizing studies and their drawbacks.

Table 4.1: Summary of the previously used methods for reducing data size for sizing studies and their drawbacks and limitations.

Method	Drawbacks or limitations	Literature
Arbitrary or unclear choice of the typical day or period	Can lead to very specific sizing for the ESS, which may be an over- or underestimation for other days.	[192], [196]–[198], [203]
Aggregation of high-resolution data into low-resolution data	Loss of high-frequency components, which can impact the sizing outcome.	[199]
Use of clustering techniques	• Requirement to specify beforehand the number of clusters. • Power profiles in one cluster are not necessarily similar. • Clusters with power profiles that are not frequent.	[194], [200]–[202]

In this thesis, a novel alternative is proposed, where reoccurring daily consumptions patterns are detected using the motif discovery algorithm [27], which are then used for the sizing of the ESS, rather than using an arbitrary day, clustering techniques or reducing the data resolution. Motif discovery uses the Symbolic Aggregated Approximation (SAX) of the data set, instead of the raw data, which helps to find similar, but yet, not the same patterns that are being repeated during the whole data set. It also uses the random projection algorithm [204], which makes the process of finding similar patterns faster than the clustering approaches [205]. Moreover, motif discovery only groups power profiles when they are very similar. There is also no need to specify the number of patterns that needs to be found, and patterns or *motifs* are generated only if they are being repeated often. The pattern with a high number of reoccurrences and high probability are then used for deriving the characteristics of the ESS, rather than using some random day or the worst-case scenario. Since the length of pattern used for the sizing data is only one day, high-resolution data can be easily employed with no concern on the computational burden of the sizing algorithm.

In this work, the motif discovery algorithm has been applied for sizing two different types of ESS at 10/0.4 kV substations for solving different challenges in the low voltage distribution grids:

- A high energy density ESS, such as a lithium-ion battery energy storage system, referred to as *type 1 ESS*, to be used for the purpose of peak shaving and load balancing. The type 1 ESS can increase the local consumption of the PV generated power, avoid the reverse power flow around noon, reduce the evening peak, and therefore, increase the PV hosting capacity of the low voltage grid.
- A high power density ESS, such as a FESS, referred to as *type 2 ESS*, to be used for the purpose power smoothing, which can limit the number and severity of rapid voltage changes at the connection point.

For deriving the characteristics of each ESS, firstly, reoccurring daily consumption patterns are detected using the motif discovery algorithm from the measurement data collected at several low voltage grids in southern Germany. Then, a low-pass filter is applied to decouple the discovered patterns into high- and low-frequency components for sizing different storage technologies. The cut-off frequency of the low-pass filter is determined through a simple optimization framework. The nominal capacity, nominal power, maximum ramp rate, and the number of times per day where each ESS changes from charging to discharging mode are calculated based on the power profiles allocated to each type of storage technology. The proposed sizing methodology is evaluated by assessing the performance of both ESS types, with the characteristics derived from the standard patterns, during the time horizon of the whole data set. In addition, the effects of data resolution, the cut-off frequency of the low-pass filter, and choosing patterns other than the most reoccurring one on the sizing outcome are also investigated. Although the proposed method is used for sizing a centralized ESS at the 10/0.4 kV substations, this method can be applied for sizing all types of ESS for any other application using historical measurement data. Examples include behind-the-meter applications of an ESS, such as increasing the self-consumption of the PV generated power or energy arbitrage.

The rest of this section is organized as follows. The collected measurement data used for deriving the storage characteristics in this work are presented in the next subsection, followed by a description of the motif discovery algorithm. The discovered standard patterns and their characteristics are presented in the following subsection. Next, the most important storage characteristics are formulated using the power profile that each ESS has to cover. The procedure to determine the optimum cut-off frequency for applying a low-pass filter on each pattern is described next with the aim optimizing the characteristics of the two major types of ESS. The sizing outcome and the effects of data resolution, cut-off frequency of the filter, and selecting other daily patterns on the storage characteristics are discussed in the following subsection. Lastly, the evaluation of the proposed sizing methodology is presented using the original full data set.

4.2.1 The Input Data for the Sizing

This work aims for installing a centralized ESS at a 10/0.4 kV substation. Therefore, the input data for the sizing study are collected at several 10/0.4 kV substations in southern Germany in the summer and fall of 2018, using the setup described in subsection 4.1.4.2. In comparison to the distributed ESS, a centralized ESS at the substation has the advantage of a single grid interface, a simpler control design, access to the electrical and monitoring systems of the substation, and availability of utility-owned land [206].

Among the substations from which data have been collected, four substations were selected for the sizing study, which represent different levels of PV penetration, and different types of loads. Moreover, the measured data has been recorded using three different resolutions of 1 second, 1 minute, and 10 minutes. Therefore, the effects of data resolution on the sizing of different storage technologies can also be investigated. A minimum of two weeks of measurement data was recorded at each substation.

As an example, the collected data with 1-minute resolution during a period of 8 days, starting from a Monday, has been illustrated in Figure 4.11 for the four selected buses. Clearly, there are patterns that are repeating themselves on a daily basis at each bus. While these patterns seem quite similar, they are also not exactly the same. For bus 1 and 2, which their power profiles are shown in Figures 4.11(a) and (b), there are different patterns occurring during the weekends (the 6th and 7th day). This is due the fact that these low voltage grids have mostly industrial and commercial loads. However, starting from the day 8, the pattern for a working day starts to reappear. In bus 1, the share of PV generation in comparison to the power consumption is so low that it can hardly be observed in the power profile during the working days, and it can only be seen during the weekends. In the power profile of bus 2, a slightly higher share of PV generation is observed, in which the increase in the consumption around noon during working days is more or less covered by the PV generation at its peak. In both cases of bus 1 and 2, negative power flow is observed only at weekends, when the load is at its minimum. Bus 3 and 4, which their power profiles are shown in Figures 4.11(c) and (d), represent two residential low voltage distribution grids with a relatively high level of PV penetration. In these cases, the consumption patterns are relatively similar for both workdays and weekends. The power profile measured at bus 3 shows a high peak only during the evening, while in the power profile of bus 4, the morning peak is as large as the evening peak. Moving from bus 1 towards bus 4, the ratio of PV generation to the peak consumption power increases, up to a point that in case of bus 3 and 4, the PV generation is dominant and leads to reverse power flows during noon in several days. Large power fluctuations around noon are also more present in buses with a higher share of PV generation, such as in bus 3, which could have been caused by passing clouds over the PV panels. It is important to note that these measurements were not conducted simultaneously, as a single measurement device was used.

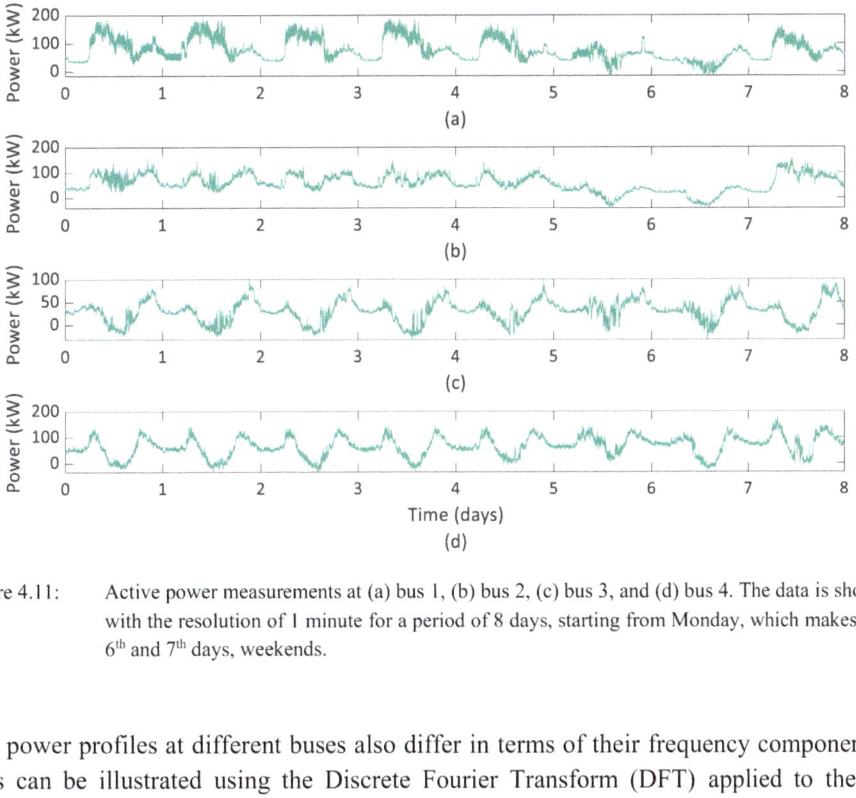

Figure 4.11: Active power measurements at (a) bus 1, (b) bus 2, (c) bus 3, and (d) bus 4. The data is shown with the resolution of 1 minute for a period of 8 days, starting from Monday, which makes the 6th and 7th days, weekends.

The power profiles at different buses also differ in terms of their frequency components. This can be illustrated using the Discrete Fourier Transform (DFT) applied to the 1-seccond measurement data. The one-sided frequency spectrum using the DFT is shown in Figure 4.12 for each bus. Each frequency component represents in this figure a periodic cycling of power. As seen, each power profile has a different amplitude at each frequency. For instance, the frequency component of 0.0116 mHz, which corresponds to every 24 hours, can be an indication of daily reoccurrence of a consumption pattern, which is highest for bus 4, followed by bus 3. This is due to having similar power profiles during the whole week, including weekends. The power profile at bus 4 shows a higher magnitude at 0.0231 mHz in comparison to other buses, which corresponds to 12 hours. This can clearly be seen from Figure 4.11(d) as well, where each day can be divided into two similar patterns. High frequency components above 2 mHz are observed mostly at bus 1, as it consists of mostly industrial loads with a high power variability.

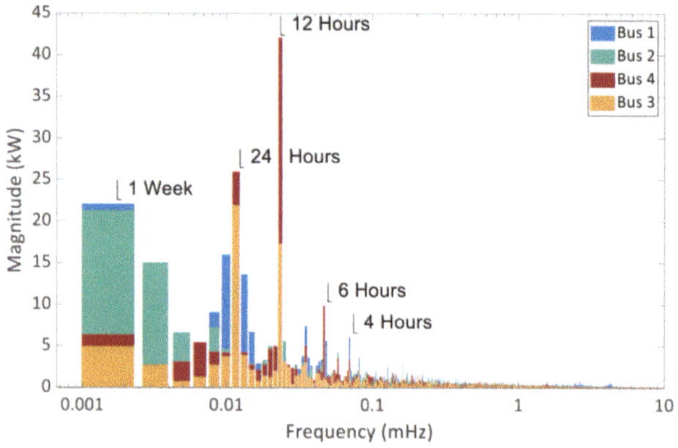

Figure 4.12: One-sided frequency spectrum of the power profiles at all buses using the 1-second data for a period of one week.

4.2.2 Introduction to Motif Discovery

As previously discussed, the first step towards sizing an ESS using historical measurement data is to select the power profile, which can accurately represent the whole data set. Instead of using an arbitrary day or the worst-case scenario, which can lead to very specific sizing outcomes, we detect and use a reoccurring daily consumption pattern at each substation using the motif discovery algorithm. The approximation of the whole time series with a standard reoccurring pattern helps us to focus on regular reoccurring patterns in the consumption, rather than on single occurrences.

Finding reoccurring patterns in time series is a common unsupervised learning problem. In this work, we propose the use of motif discovery, introduced in [27], and its improved version in [207], tailored for energy-related time series. A major advantage of using motif discovery is the use of the Symbolic Aggregated Approximation (SAX) of the time series rather than the time series itself for finding similar patterns. This allows to find patterns that are similar and yet, not the same. In contrast to the established clustering approaches [194], [200]–[202], there is also no need to specify beforehand the number of groups or clusters, and power profiles are only grouped together, when they are found to be similar. Power profiles that are rarely reappearing are automatically left outside.

Motif discovery has been applied in various disciplines for a variety of applications, including identifying seismic activity patterns in active volcanos [208] and classifying heartbeat sounds [209]. In the realm of energy, motif discovery has been applied for finding flexibility potentials in industrial processes [210] and non-intrusive load monitoring [211]. Prior to this work, there are no reports on using motif discovery for storage sizing studies or other applications in power systems.

The algorithm consists of five main steps, as shown in Figure 4.13, which are described in the following according to [207]. The term *motif* refers to a pattern or a series of sequences in the time series that are repeated often.

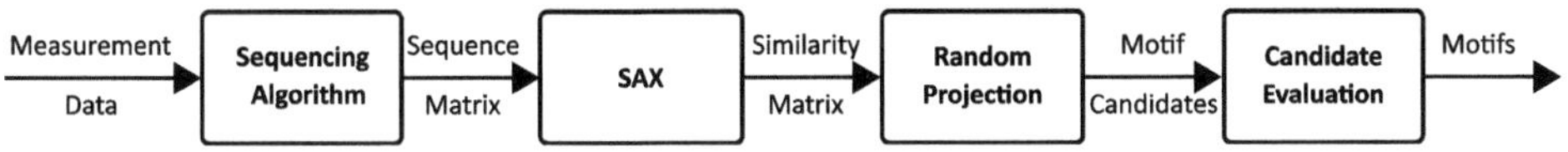

Figure 4.13: The process of the motif discovery using SAX and random projection (adopted from [207]).

Sequencing algorithms: In the first step, the whole time series data, which consists of n data points, is divided into sequences with equal length of ϖ. As we are looking for daily patterns in this work, the length of the sequence is set to correspond to 24 hours, depending on the data resolution. For instance, for the 1-second data, ϖ is 86400.

Symbolic Aggregated Approximation (SAX): Next, the sequences are converted into a series of letters out of the alphabet, similar to the DNA sequences, from which this method was inspired. To do so, the sequence time series Y is divided into equal pieces with a predefined length of p. All data points within each piece are aggregated by their mean value, $y_{\mathrm{i}}^{\mathrm{PAA}}$. This is known as Piecewise Aggregated Approximation (PAA) of the data, which helps to reduce the effect of noise and minor variations in finding patterns.

In the following step, a letter out of the alphabet is assigned to each piece, according the value of $y_{\mathrm{i}}^{\mathrm{PAA}}$. First, a section of the alphabet with the predefined length of a together with a list of breaking points β are defined, in which $\beta = \beta_1, \beta_2, \dots, \beta_{\mathrm{a}-1}$, and $\beta_{\mathrm{i}-1} < \beta_{\mathrm{i}}$. The breaking points β are chosen to be the quantiles of the empirical cumulative distribution function of the sequence, which divides the area underneath into a equal areas. We assign $a = 10$, so that each letter corresponds to a meaningful quantile of the distribution function. If the mean value of each peace ($y_{\mathrm{i}}^{\mathrm{PAA}}$) is in the range of $[\beta_{\mathrm{j}-1}, \beta_{\mathrm{j}})$, then its SAX representation ($y_{\mathrm{i}}^{\mathrm{SAX}}$) is the corresponding letter of α_{j}, or

$$y_{\mathrm{i}}^{\mathrm{SAX}} = \alpha_{\mathrm{j}} \qquad \text{if, } y_{\mathrm{i}}^{\mathrm{PAA}} \in [\beta_{\mathrm{j}-1}, \beta_{\mathrm{j}}). \tag{4.15}$$

For observations $x = (x_1, x_2, \dots, x_n)$, the empirical cumulative distribution function $F_n(p)$ is the fraction of observations which are less than or equal to p, i.e.,

$$F_n(p) = \frac{1}{n}\sum_{i=1}^{n} I(x_i \leq p). \tag{4.16}$$

At the end of this step, we have strings of letters or *words* representing each sequence, which in our use case are the daily power profiles. The PAA and SAX of a time series are illustrated more intuitively in Figure 4.14. In this figure, the original time series is shown in grey, the black line shows the PAA of the data, and the blue letters show its SAX representation.

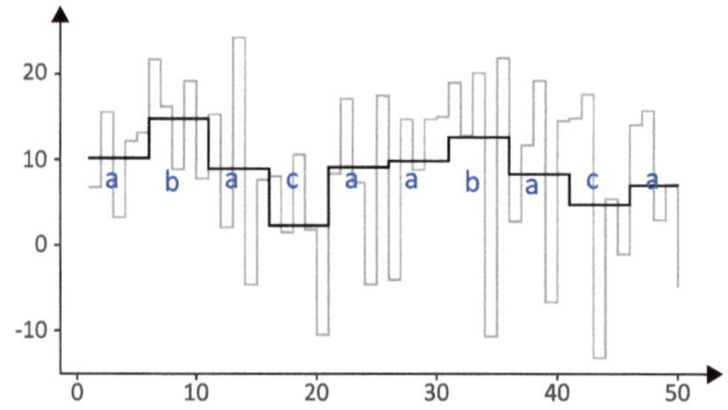

Figure 4.14: An example of PAA of a time series (black line) and its SAX representation (the blue letters). The time series itself is shown with the grey line (adopted from [207]).

Random projection: To find the potential motifs, one can simply compare each sequence to all other sequences to look for similarities, but this can be time consuming for large data sets. Instead, motif discovery uses random projection [204], which converges much faster. The SAX representations of the data are saved in a similarity matrix $S \in (n - m + 1) \times \varpi$ as rows, using a sliding window. Here, n is the number of observations, and m is the number of sequences. In each iteration of the random projection algorithm, k columns of the similarity matrix S are selected randomly, where $k \leq \varpi$, and k is a user-defined parameter. The words that are built with the selected k letters are compared with all rows of S. When the words are similar, the corresponding entry in the so-called collision matrix is incremented. The entries with the highest values in the collision matrix are considered potential motifs. The motifs are then transformed back to their original data form, which are the daily power profiles in this use case.

Candidate evaluation: The aim of this final step to find how often the candidate motifs occur during the whole time series. The potential motifs are iterated over the original time series to find the instances where the motifs occur, which are the days that follow the same pattern. Therefore, while finding motifs is based on their SAX representation, their evaluation is carried out using the original time series.

Since the length of the sequences are equal in this use case, a simple Euclidean distance measure is used to find the occurrences of each possible motifs. For two sequences of y_i and y_j, $d(y_i, y_j)$ is the Euclidean distance between them, and is calculated according to

$$d(y_i, y_j) = \sqrt{\sum_{t=1}^{n} \left(y_i(t) - y_j(t)\right)^2}. \tag{4.17}$$

Each sequence similar to a motif is considered an occurrence of the motif. Each motif can have several occurrences within the whole time series. We define the *standard pattern* as the 80 %-quantile of all the occurrences of the motif with the highest number of occurrences. This quantile is chosen, as the loads are also supported by the grid in this use case. Choosing a higher percentage quantile is also possible, which will lead to larger storage units.

4.2.3 Standard Patterns for Sizing ESS

Using the described motif discovery algorithm, the standard consumption patterns at the four low voltage distribution grids are obtained. A minimum of two weeks of measurement data have been used to derive these patterns. However, larger data sets can also be easily used, considering the fast convergence of the random projection algorithm. The standard patterns are derived separately for the 1-second, 1-minute, and 10-minute data for each low voltage distribution grid, in order to study the effect of data aggregation on the sizing outcome.

The standard patterns using the 1-minute data for all buses are shown in Figure 4.15. Similar patterns are derived using the data with other resolutions. As seen in Figure 4.15(a), the standard pattern of bus 1 represents a low voltage grid with small-scale industrial facility, which is active only within working hours, and has a low share of PV generation. The share of PV generation at bus 2 is slightly higher, and just enough to reduce the midday peak to the base load value, while for the buses 3 and 4 the PV generation can almost cover the load for short periods at its peak. The standard pattern at bus 4 shows a slight reverse power flow between 13:00 and 14:00.

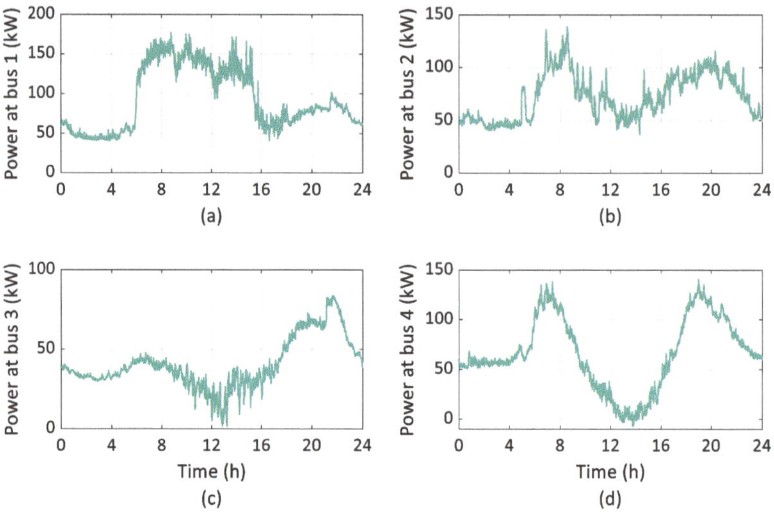

Figure 4.15: The standard consumption patterns at the four buses: (a) Bus 1: No PV generation with industrial loads. (b) Bus 2: Limited PV generation to cover only the increase in consumption at peak generation. (c) Bus 3: Significant share of PV generation in a residential grid, resulting in a typical *duck curve*. (d) Bus 4: High share of PV production in a residential grid, leading to slight negative power flow at peak generation.

In order to see the effect of PV generation on the frequency components of the power profiles, a time-frequency analysis on the standard patterns is carried out. Figure 4.16 shows the spectrogram of the standard patterns using the 1-second data of the four buses, which is calculated using the short-time Fourier transform and a Hamming window with the length of 1-minute with no overlaps. The spectrum ends at half of the sampling frequency, since according to the Nyquist principle, the spectrum is symmetrical around this value. As seen in Figures 4.16(c) and (d), at bus 3 and 4 with a high share of PV generation, the frequency content varies significantly with the increase in the PV generation around noon at both high and low frequencies. The constant power lines on the spectrum become vertical from being horizontal, which indicates that the PV generation alters a vast range of frequencies at the same time. With the decrease in PV generation at the evening, the frequency spectrum is restored, and shows more or less the same frequency components until the end of the day.

Another interesting observation is that at bus 1, which supports an industrial facility, both high- and low-frequency components show high values within the activity period of the industrial facility, with the exception of the lunch break at approximately 12:00, as seen in Figure 4.16(a).

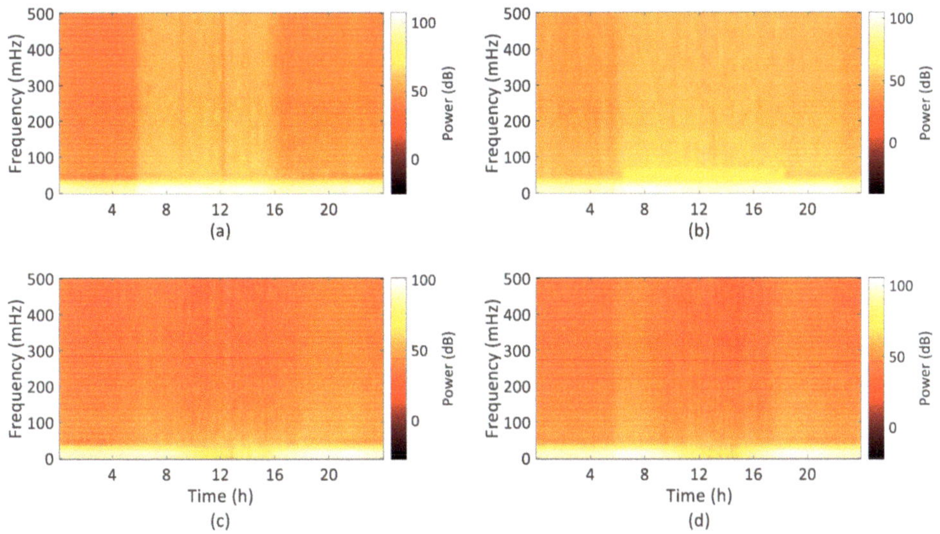

Figure 4.16: The spectrogram of the standard patterns using the 1-second data and hamming window of 1-minute length with no overlaps for (a) bus 1, (b) bus 2, (3) bus 3, and (d) bus 4.

The standard patterns of Figure 4.15 are the motifs with the highest number of reoccurrences. However, it is possible that motif discovery algorithm detects several other patterns as well, with lower number of occurrences. This can be the case, for instance, where workdays and weekends follow different consumption patterns, or large data sets such as yearly time series are used, where seasonal power profiles can be quite different. In our data sets, several patterns were detected at bus 2 when using the 1-minute and 10-minute data. When using the 1-minute data, three motifs have been detected at bus 2, which are shown in Figure 4.17. In this figure, pattern 1 represents a normal weekday pattern, which has the highest number of occurrences, and it is the main pattern used for the storage sizing later within this chapter. Patterns 2 shows the consumption pattern during the weekends and public holidays, while patterns 3 represents the power profile of the first working day after a weekend or a holiday. The difference between pattern 1 and 3 can be associated with the loads that stay connected during the workdays but are disconnected for weekends. The effect of using the other patterns with lower number of reoccurrences on the sizing is investigated later in subsection 4.2.6.3, which can also show how an arbitrary choice of days for a sizing study can lead to different sizing outcomes.

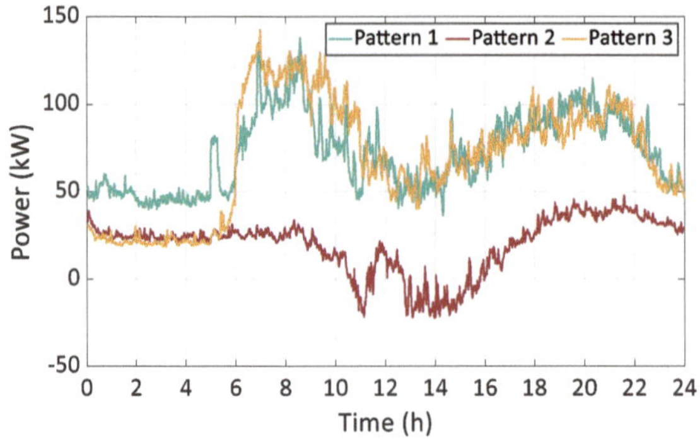

Figure 4.17: The three different patterns detected at bus 2 using motif discovery. Pattern 1 is the consumption pattern during a typical workday with the highest number of reoccurrences, while pattern 2 shows the consumption pattern over the weekdays and public holidays. Pattern 3 represents the power profile of the first day after a weekend or a holiday.

4.2.4 Deriving Energy Storage Characteristics

The standard power profile at each bus is used to calculate the required ESS characteristics. Here, it is assumed that the daily average of the standard patterns is covered by the grid and the ESS provides the variations from the daily average. Using this approach, the evening peak, the reverse power flow, the high ramp rate requirement during the evening hours, and the high-frequency power fluctuations are reduced using the ESS, and the distribution system operator sees a roughly constant demand from the low voltage grid throughout the day.

But different storage technologies operate at different time scales. They also differ substantially in terms of their lifetime and the maximum number of useful cycles. As discussed in Chapter 2, while a FESS can provide a significant number of short-term power cycles with no degradation over its lifetime, a high energy density ESS, such as a lithium-ion battery, are more limited in this regard. Therefore, in our use case, it is suggested to use a high energy density ESS for peak shaving and load balancing, while a high power density ESS such as a FESS is used to cover the short-term power fluctuations. This helps smoothing the charging and discharging profile of the lithium-ion battery, and therefore, preserving its lifetime and capacity [55]. Both flywheels and supercapacitors can be used for smoothing the charging profile of a battery, but it has been recently demonstrated that a hybrid battery-flywheel system outperforms a battery-supercapacitor combination in terms of reducing the peak current drawn from the battery, and the overall system efficiency [57]. Reducing the maximum current drawn from the lithium-ion cells can limit the degradation of the anode active material and lithium plating, as shown in [17].

Therefore, the low-frequency components of the variations in power from its average are allocated to a high energy density ESS with limited number of cycles, referred to as *type 1 ESS*. The high-frequency components are allocated to a high-power density ESS, referred to as *type 2 ESS*, for smoothing out the short-term power variations. Each storage technology can also be used individually, if required. The type 2 ESS, such as a FESS, can be used alone to smoothen the sharp active variations drawn from the grid in order to avoid fast voltage changes in weak parts of the grid, similar to the use case of section 4.1. It can also be used in combination with a type 1 ESS in order to extend its lifetime. The type 1 ESS can also be installed alone to reduce the evening peak, avoid reserved power, and reduce the required ramp rate during evenings. However, in this case, it also has to cover the short-term power fluctuations, which can reduce its effective lifetime. Combining these technologies in a hybrid ESS can provide all these advantages at once.

The following describes how the nominal capacity, the nominal power, the maximum ramp rate, and the number of mode changes from charging to discharging and vice versa, are derived from the allocated power profile to each ESS. These calculations are repeated for each type of ESS, each data resolution, and each standard pattern. Practical aspects such as the energy conversion losses and the requirements to preserve the lifetime of the type 1 ESS are also considered in the sizing calculations.

4.2.4.1 Nominal Capacity

The nominal capacity of an ESS determines the characteristics of the storage medium, such as the number of cells in a lithium-ion battery, or the inertia and maximum rotational speed in a FESS, as discussed in Chapter 2.

The nominal capacity ($\mathrm{E_n}$) can be calculated by integrating over the daily power profile allocated to each ESS, and then computing the difference between the highest and lowest peaks of the output signal. The efficiency of the ESS (η), which includes the efficiency the power electronics interface and the storage medium, should also be considered, as the ESS should provide higher power than the allocated power while discharging, and lower power is stored in the ESS while charging, considering the conversion losses. The sign function (sgn) is used to determine from the power profile that whether the ESS is being charged or discharged for applying the efficiency of the ESS.

In addition, when using storage technologies such as lithium-ion batteries, it can be beneficial to the system's lifetime to avoid very low and high values of state of charge. It has been shown for lithium-ion batteries that avoiding high values of state of charge (above 80%) can significantly help reduce the cathode degradation and calendar aging of the battery [17]. Similarly, deep discharging should also be avoided, as it can significantly increase the internal resistance in the lithium-ion cells. Therefore, it is recommended to reserve a certain non-usable capacity to prevent deep discharging and over-discharging the system [17]. Hence, in this work, the state of charge is kept within the certain range of $[\mathrm{SOC_{min}}, \mathrm{SOC_{max}}]$ in order to preserve the system lifetime. Considering all these factors, the nominal capacity of the ESS is calculated using

$$\mathrm{E_n} = \frac{max\left(\eta^{-sgn(P(\tau))} \int_0^t P(\tau)d\tau\right) - min\left(\eta^{-sgn(P(\tau))} \int_0^t P(\tau)d\tau\right)}{\mathrm{SOC_{max}} - \mathrm{SOC_{min}}}. \tag{4.18}$$

In Eq. (4.18), $P(\tau)$ is the allocated power profile to the ESS, and t corresponds to 24 hours, depending on the data resolution.

For the type 1 ESS, the parameters for a lithium-ion battery are assumed. The minimum state of charge is assumed to be 10 %, while the maximum state of charge is chosen to be 80 %, as recommended in [17]. The roundtrip efficiency is assumed to be 90 %, which is the efficiency of a Tesla PowerWall 2 home battery system [212]. The same efficiency is used for the type 2 ESS, with the minimum and maximum values for the state of charge to be 25 % and 100 %, respectively. This is due to the fact that a FESS have no constraints on being fully charged, but the machine's torque constraint limits the maximum deliverable power at low values of state of charge (see subsection 6.3.1).

4.2.4.2 Nominal Power

The nominal power ($\mathrm{P_n}$) of an ESS is often limited by its power electronics interface. It is simply the maximum value of the power profile allocated to the ESS, considering both charging and discharging modes, and the efficiency of the ESS. Therefore,

$$\mathrm{P_n} = \max\left(\left|\eta^{-sgn(P(\tau))} P(\tau)\right|\right). \tag{4.19}$$

4.2.4.3 Maximum Ramp Rate

In this work, the maximum ramp rate ($\mathrm{R_n}$) is considered to be the maximum change in the storage power between two consecutive measurement points in time. Considering the sampling frequency (f_s) of the measured power signal, the maximum ramp rate is calculated as

$$\mathrm{R_n} = max\left(\eta^{-sgn(P(\tau))} P(\tau) - \eta^{-sgn\left(P\left(\tau - \frac{1}{f_s}\right)\right)} P\left(\tau - \frac{1}{f_s}\right)\right). \tag{4.20}$$

4.2.4.4 Number of Mode Changes and Lifetime

Since an ESS is not fully charged and discharged with every cycle, the number of times that each ESS changes from charging to discharging mode in each day is calculated in this work, rather than calculating the number of full cycles. The number of mode changes is simply the number of zero crossings of allocated power profile the ESS ($P(\tau)$).

Since the lifetime and the number of cycles is limited for lithium-ion batteries, several contributing factors to the aging this type of ESS are considered, including [17], [199]:

1. Operating and maintaining the system at very low and very high values of state of charge: This is avoided by simply selecting a larger storage capacity, according to Eq. (4.19), and staying within the given range of state of charge.
2. Frequent variations of rate of charge or discharge: This is avoided by using a low-pass filter and allocating the high-frequency components to another type of ESS, such as a FESS.
3. Avoiding high rates of charge and discharge current: The power required from the lithium-ion battery is reduced by using a second type of ESS such as a FESS for providing the short-term peaks.
4. Charging or discharging more than once per day, which reduces the calendar life of the system: This is avoided by ensuring that the ESS is only charged and discharged fully only once a day.

4.2.5 Allocating Power Profiles

The use of a low-pass filter to separate high- and low-frequency components of the variations in the active power has become a standard approach for a simultaneous sizing of several storage technologies. This is particularly the case for sizing a battery-supercapacitor hybrid ESS, where the supercapacitor can increase the power capability of the battery, and preserve its lifetime [213], [214]. The choice of the cut-off frequency of the filter can significantly impact the sizing outcome [197], [214], which can be any value in the range of $(0, f_s/2)$, in which f_s is the sampling frequency of the measured data. Typical discharge times of each storage technology [215], market intervals [216], and optimization frameworks [192] have been used for selecting the filter's cut-off frequency. However, there is no generalized approach for selecting the cut-off frequency, which can be applied to all applications. In our use case, a relatively low cut-off frequency allocates most high-frequency components to the type 2 ESS, but also increases its nominal capacity, which can be quite costly considering the low energy density of these technologies. With a relatively high cut-off frequency, the type 1 ESS has to cover many fast variations in the power profile, which limits the advantages gained by using the second ESS. In this work, a simple optimization framework is introduced for selecting the cut-off frequency, which aims to simultaneously minimize the capacity of the high power density ESS, while reducing the maximum ramp rate and power of the high energy density ESS. This optimization is based on the following assumptions:

1. Increasing the capacity of type 2 ESS, e.g., flywheels or supercapacitors, has higher costs than increasing the capacity of type 1 ESS, e.g., batteries.
2. Due to the higher power density of the type 2 ESS, it has a smaller footprint for providing high powers than the type 1 ESS, and therefore, a higher rated power for the type 2 ESS is preferred.
3. It is desired to have a low ramp rate for the type 1 ESS for a prolonged lifetime, while type 2 ESS has no limitations in this regard.

With the above-mentioned assumptions, a simple objective function is defined as

$$minimize \quad Obj = \mathrm{k}_1 \frac{\frac{E_{\mathrm{n2}}}{E_{\mathrm{n1}}}}{\max\left(\frac{E_{\mathrm{n2}}}{E_{\mathrm{n1}}}\right)} + \mathrm{k}_2 \frac{\frac{P_{\mathrm{n1}}}{P_{\mathrm{n2}}}}{\max\left(\frac{P_{\mathrm{n1}}}{P_{\mathrm{n2}}}\right)} + \mathrm{k}_3 \frac{\frac{R_{\mathrm{n1}}}{R_{\mathrm{n2}}}}{\max\left(\frac{R_{\mathrm{n1}}}{R_{\mathrm{n2}}}\right)}. \tag{4.21}$$

In Eq. (4.21), the nominal capacity, the nominal power, and the maximum ramp rate are calculated according to Eq. (4.18)-(4.20) for each type of ESS, which are denoted with 1 and 2 for the type 1 and type 2 ESS, respectively. Also, k_1, k_2, and k_3 are weighting coefficients, which are assumed to be equal to one in this work, but can be adjusted when required. The objective function of Eq. (4.21) aims to minimize simultaneously the ratio of capacity of the type 2 the ESS to the capacity of the type 1 ESS, the ratio of the power of the type 1 ESS to the power of the type 2 ESS, and the ratio of the maximum ramp rate of the type 1 ESS to the one of the type 2 ESS. All these three factors are normalized using their maximum value within the range $(0, f_s/2)$. To find the optimum cut-off frequency, a simple gradient search method is used, similar to [197], in which the cut-off frequency is increased from its minimum value up to the point that is no longer decreasing. The frequency, at which the objective function is at its minimum is used as the cut-off frequency of the filter for decoupling the standard patterns. These values are listed in Table 4.2 for the standard pattern of each bus using the 1-second and 1-minute data resolutions. As the discharge time of the type 2 ESS is often much shorter than 10 minutes, their characteristics is not calculated using the 10-minute data.

The optimum cut-off frequencies shown in Table 4.2 are used to decompose the standard patterns into high- and low-frequency components and to allocate these power profiles to each ESS type. Using this approach, the nominal capacity of type 2 ESS is minimized, while simultaneously reducing the power and the maximum ramp rate of type 1 ESS for a prolonged lifetime. Figure 4.18 depicts the change in the objection function of Eq. (4.21) for each bus with the increase in the cut-off frequency of the filter.

As discussed, using the low-pass filter, the high-frequency components of the active power variations are covered by a type 2 ESS. As an example, Figure 4.19 shows the decomposed power profile used for sizing each type of ESS at bus 1. As seen, the short-term power peaks, for instance, the 40-kW short pulsed power after 8 a.m., are covered by the type 2 ESS, while the type 1 ESS covers only the slower variations in power.

Table 4.2: The optimum cut-off frequency of the low-pass filter for each bus using different data resolutions.

	Data resolution	
Bus	**1 second**	**1 minute**
Bus 1	59 mHz	3.4 mHz
Bus 2	56 mHz	3.2 mHz
Bus 3	50 mHz	3.5 mHz
Bus 4	46 mHz	3.2 mHz

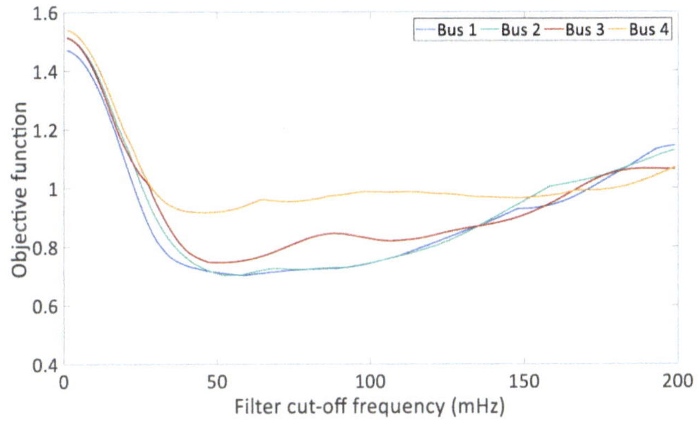

Figure 4.18: Changes in the objective function with the increase of the cut-off frequency of the filter at each bus using the data with 1-second resolution.

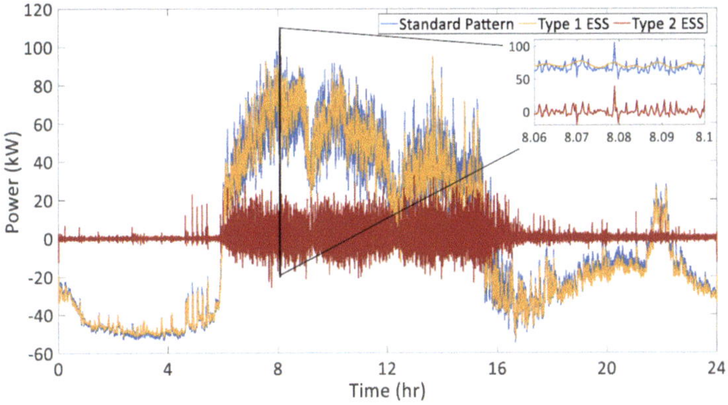

Figure 4.19: The allocated power to each type of ESS at bus 1 by applying a low-pass filter with the optimum cut-off frequency.

4.2.6 Sizing Outcome

Using the standard patterns of each bus and the allocated power profile to each ESS by applying the low-pass filter with the optimum cut-off frequency, the ESS characteristics has been calculated for each bus and each data resolution from Eq. (4.18)-(4.20). The results are shown in detail in Table 4.3.

Table 4.3: Sizing results according to the standard patterns at each bus for the three different data resolutions.

Bus	ESS Type	Data Resolution	Capacity (kWh)	Nominal power (kW)	Max. ramp rate (kW/s)	Mode changes per day
1	Type 1	1 s	615.32	94.87	4.84	34
		1 min	632.09	84	0.31	8
		10 min	629.59	79.42	0.07	2
	Type 2	1 s	1.99	38.94	34.92	28500
		1 min	2.47	20.75	0.46	813
2	Type 1	1 s	268.94	63.82	4.29	584
		1 min	219.23	70.09	0.26	46
		10 min	394.79	46.22	0.04	10
	Type 2	1 s	0.94	24.64	23.50	30352
		1 min	0.67	8.13	0.21	848
3	Type 1	1 s	268.53	56.96	1.46	226
		1 min	212.64	46.35	0.13	22
		10 min	257.69	50.94	0.03	6
	Type 2	1 s	0.28	9.35	9.69	24780
		1 min	0.70	7.59	0.17	895
4	Type 1	1 s	450.5	87.34	1.99	106
		1 min	421.20	76.32	0.12	12
		10 min	434.10	73.05	0.04	6
	Type 2	1 s	0.38	11.41	14.01	22415
		1 min	0.90	8.96	0.25	588

Considering the sizing outcome for bus 1 using the 1-second data as an example, the rated capacity and power for the type 1 ESS is calculated to be 615.12 kWh and 94.87 kW, respectively. For the type 2 ESS with a high power density (e.g., a FESS), the nominal capacity and power are estimated to be 2 kWh and 39 kW. More importantly, using the type 2 ESS, the maximum ramp rate of type 1 ESS is calculated to by only 4.84 kW/s, while the type 2 ESS has to provide a ramp rate of around 35 kW/s. Without the type 2 ESS, this high ramp rate has to be covered by the type 1 ESS. Moreover, the number of mode changes, where the ESS changes from charging to discharging mode is significantly reduced for the type 1 ESS, as most of the small variations in power are covered by the type 2 ESS. Similar results are obtained for other buses, as seen in Table 4.3. The maximum ramp rate and the number of mode changes for the four buses using the 1-second data are also plotted in Figure 4.20 for an easier comparison.

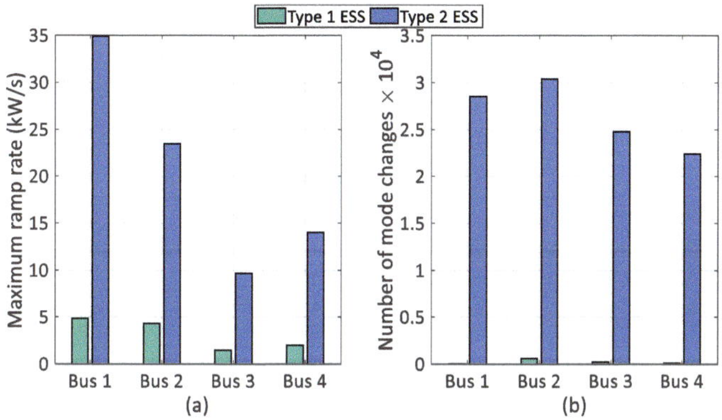

Figure 4.20: (a) The maximum ramp rate, and (b) the number of mode changes per day for both types of ESS using the 1-second data.

For buses 2-4, a type 2 ESS with capacity of less than 1 kWh is required for smoothing out the power profile of type 1 ESS. When comparing the results of the four buses, it can be noted that in buses 3 and 4, which are residential areas with a high share of PV generation, the type 2 ESS requires less ramping rate and a smaller number of mode changes in comparison to bus 1, where industrial loads are connected. Therefore, it can be seen that the variability of industrial loads such as the ones connected at bus 1 can be more demanding for an ESS, in comparison to the variability of PV generated power.

In the following the effect of data resolution, the cut-off frequency of the low-pass filter, and selecting patterns other than the most reoccurring one on the sizing outcome is investigated.

4.2.6.1 The Effect of Data Resolution on the Sizing

As shown in Table 4.3, the ESS characteristics has been calculated using the data with 1-second, 1-minute, and 10-minute resolutions. It can be observed that the nominal power and capacity of a type 1 ESS are more or less in the same range in most cases, when lower data resolutions are used. The exception is using the 10-minute data for bus 2, since the pattern detected using the 10-minute data is quite different from the pattern detected using other data resolutions. Therefore, it can be concluded that low resolution data up to 10-minutes does not necessarily lead to large errors in estimating the capacity and power of type 1 ESS such as a lithium-ion battery, for the application of peak shaving and balancing the load and PV generation. However, the maximum ramp rate and the number of times the ESS changes from charging to discharging mode and vice versa, cannot be accurately estimated when low data resolutions are used. Hence, using high resolution data, such as 1-second data, should be preferred, when available.

For sizing the type 2 ESS for power smoothing applications, it is necessary to use data with the highest resolution. Even using 1-minute data can lead to extreme over-sizing of the ESS, as in the case of buses 3 and 4, where the estimated capacity using the 1-minute data is twice the estimated value using the 1-second data. Similar to type 1 ESS, using low resolution data leads to a large underestimation of the required ramp rate and number of mode changes per day for the type 2 ESS.

4.2.6.2 The Effect of Filter's Cut-off Frequency

The sizing result of Table 4.3 were based on the optimum cut-off frequency calculated from the objective function of Eq. (4.21), and given in Table 4.2. Figure 4.21 shows how the characteristics of the type 1 and 2 ESS alters for bus 2, as an example, if other values are used for the cut-off frequency of the low-pass filter.

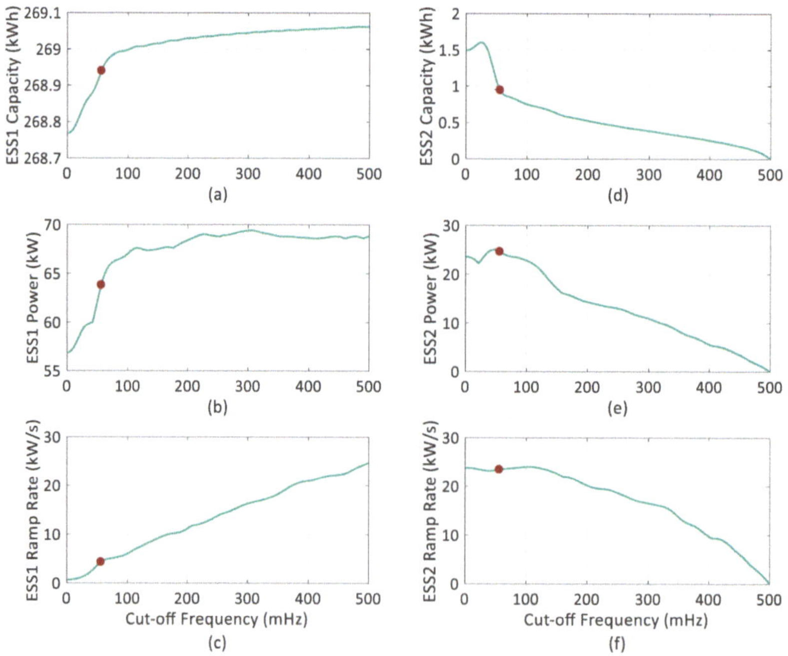

Figure 4.21: The variations in the characteristics of the type 1 and type 2 ESS with respect to the change in the cut-off frequency of the low-pass filter for bus 2. The optimum frequency is the red dot.

As seen from Figure 4.21, with an increase in the filter's cut-off frequency, the power and the maximum ramp rate of the type 1 ESS increases, while its capacity remains relatively unchanged. On the contrary, the capacity and power of type 2 ESS decreases drastically with the raising cut-off frequency, as less power is allocated to this storage technology, up to reaching the Nyquist frequency, where all frequency components are allocated to the type 1 ESS. The optimum cut-off frequency, shown with the red marker in Figure 4.21, is the point where the capacity of the type 2 ESS is relatively low (below 1 kWh), while simultaneously having a relatively low ramp rate and power demanded from the type 1 ESS.

4.2.6.3 The Effect of Selecting Different Patterns

As shown in Figure 4.17, the motif discovery algorithm can possibly generate several different patterns, which are also reoccurring, but with lower number of reoccurrences. Table 4.4 shows the sizing outcome for bus 2 using the 1-minute data, if other patterns are selected. As discussed before and shown in Figure 4.17, pattern 2 is the pattern of a weekend, while pattern 3 corresponds to the first day after a weekend or holiday.

Table 4.4: Sizing results using three different standard at bus 2 using the 1-minute data.

Bus	ESS Type	Pattern	Capacity (kWh)	Nominal power (kW)	Max. ramp rate (kW/s)	Mode changes per day
2	Type 1	Pattern 1	219.23	70.09	0.26	46
		Pattern 2	207.44	41.92	0.13	18
		Pattern 3	408.87	74.33	0.17	28
	Type 2	Pattern 1	0.67	8.13	0.21	858
		Pattern 2	0.60	9.61	0.30	826
		Pattern 3	0.72	10.80	0.27	883

It can be perceived from Table 4.4 that selecting different patterns can lead to different sizing outcomes for the type 1 ESS. When using pattern 2, which is reoccurring pattern during the weekends or holidays, the nominal power of the type 1 ESS is largely underestimated. When pattern 3 is used, which is the pattern of the first working day of the week, the calculated nominal capacity is significantly higher, due to the different power requirement during the first 6 hours of the day (see Figure 4.17). This shows how a random and arbitrary choice of the day for the sizing study, which indeed could be a weekend or a first day of the week, can lead to very different results, in particular for sizing the type 1 ESS. Therefore, such an approach should be avoided.

For the type 2 ESS, the sizing outcomes are not considerably different, when other patterns are used. This is due the fact that the high-frequency components above the filter's cut-off frequency have been more or less consistent in all patterns in our use case.

4.2.7 Evaluation of the Proposed Sizing Method

In order to evaluate the proposed sizing approach using standard patterns, the ESS with the characteristics derived from the standard patterns are tested throughout the whole data set. Since not all days follow exactly the same pattern, and the 80%-quantile of the most reoccurring pattern is selected for the storage sizing, it is evident that the ESS with the characteristics based on the standard pattern will not be able to fulfil its purposes at all times and during all days. However, since the ESS is sized based on the daily pattern that has been repeated the most, it should be capable to fulfil its purposes for most of the days within the data set. The performance of the ESS is evaluated based on how well the peak shaving and power smoothing goals can be fulfilled using the two types of ESS throughout the whole measurement period.

In order to see the effect of using type 2 ESS, such as a FESS, for power smoothing applications, the results are shown for the 1-second data only. Also, bus 3 is selected for demonstrating the performance of the storage systems, as it represents a typical duck curve with a high share of PV generation. Similar results can be shown for the other buses and using other data resolutions. Figure 4.22 shows the active power drawn from the medium voltage grid at bus 3 during the period of two weeks, which is the length of the original data set used for detecting patterns.

The results are shown for three different cases:

- **Case 1:** No ESS is installed.
- **Case 2:** Only a type 1 ESS, for example, a lithium-ion BESS, is installed for the application of peak shaving and balancing the load and PV generation.
- **Case 3:** The type 1 ESS is complemented by a type 2 ESS such as a flywheel in a hybrid structure for power smoothing applications.

Within the two weeks of measurement data, the capacity, nominal power, and the maximum ramp rate calculated from the standard patterns are used as the constraints for charging and discharging of both types of ESS. Also, the state of charge of both types of storage systems are maintained within the range of the state of charge discussed in subsection 4.2.4, which is also illustrated in Figure 4.23. A simple on/off controller is used for the peak shaving application using the type 1 ESS, which stores the PV generation starting from 10:00 until the ESS is fully charged, and discharges it from 18:00, until the storage is empty. These times are also selected based on the detected standard pattern at this bus. For the type 2 ESS, a continuous operation is considered in order to continuously filter out short-term power variations.

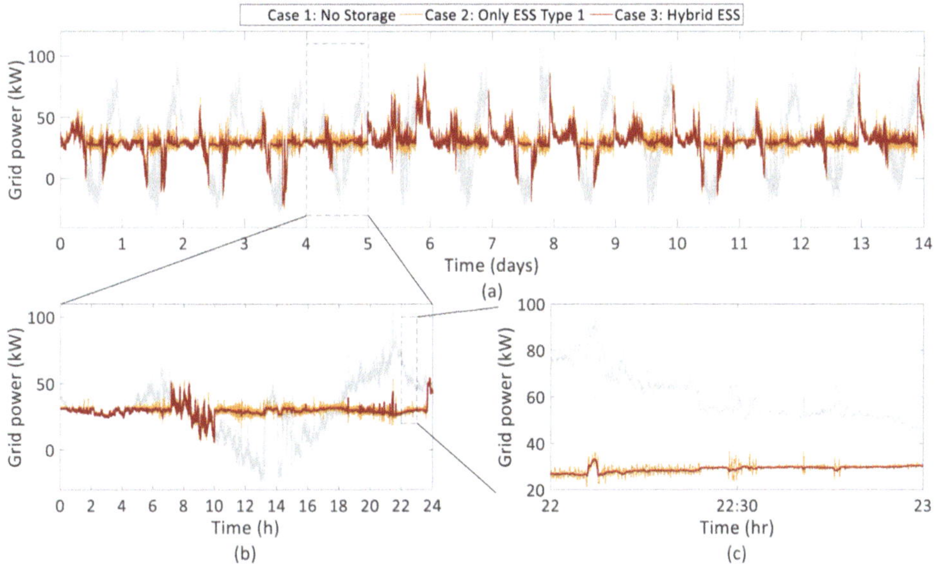

Figure 4.22: The active power drawn from the medium voltage grid at bus 3, with and without the type 1 ESS, and in combination with a type 2 ESS, such as a FESS, (a) during the whole two weeks, (b) during the 5th day, and (c) between 22:00 and 23:00 of the 5th day.

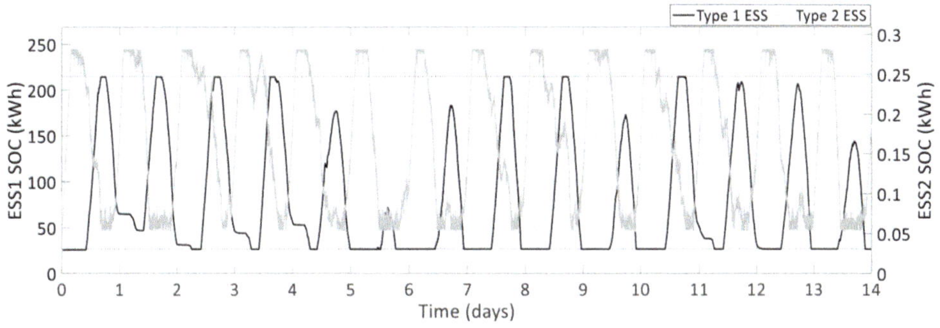

Figure 4.23: The state of charge variations during the two weeks for both ESS types, with the characteristics derived from the standard patterns.

From Figure 4.22, it is evident that:

1. In almost all the days, the evening peak is either eliminated (in 8 out of 14 days) or at least reduced by using a type 1 ESS. The exception is the 6th day, in which the PV generation was not high enough to charge the type 1 ESS and cover the demand in the evening (see Figure 4.23). Similar to any peak shaving application using an ESS, this has the potential to defer grid expansions or relief congestions.
2. The use of the type ESS 1 flattens the daily load profile at this low voltage grid. Therefore, it no longer shows a typical duck curve. This means that it is not necessary for the grid during most of the days to support the evening demand with a relatively high ramp rate. However, when the PV generation is not high enough, the type 1 ESS is fully discharged while the demand is still high. In this case, the grid has to cover the instantaneous increase in the demand.
3. The reverse power flow during noon is either fully avoided (for 8 out of 14 days) or its duration and amplitude is notably reduced. This can potentially increase the hosting capacity of the low voltage grid for new PV installations or avoid power curtailments of the PV generation due to over-voltage incidents. However, this assumption is only valid if the PV systems are installed near the ESS installation location. For PV systems that are further away from the ESS, local measures may be still necessary to avoid over-voltages.
4. With the use of the type 2 ESS, short-term power fluctuations, which can lead to fast voltage changes, are avoided.
5. By applying the low-pass filter and using the type 2 ESS, the charging and discharging profile of the type 1 ESS is smoothened, and its required ramp rate is significantly reduced, which can potentially extend its lifetime. Without the type 2 ESS, the fast active power changes should be either covered by the type 1 ESS, which can affect its lifetime, or by the grid, which can lead to fast voltage changes.

It can be concluded, that each ESS with the characteristics derived from the standard patterns can fulfil its purposes for most of the days within the data set, showing the effectiveness of the proposed method.

4.3 Summary

In this chapter, two novel data-driven solutions were introduced for the problem of allocation and sizing of an ESS.

The first section presented an allocation algorithm for a FESS for power smoothing applications using the concept of mutual information. A data-driven method based on the concept of mutual information was proposed for estimating the relative voltage sensitivity to active and reactive power changes. Unlike the classical linearized approach, the proposed approach does not require a grid model, and can capture system nonlinearities and the changes in the system operation point. The suggested method was applied for the allocation of a FESS in a real grid in southern Germany. By collecting and using measured data at the most attractive candidates, it has been shown that the proposed approach can successfully find the locations with a high voltage sensitivity to active changes. These locations can be suitable for the installation of a FESS, if sharp active power variations, for example, caused by pulsed power loads, are also present.

In the second section, a new data-driven method is introduced for sizing different types of ESS, including a high power density ESS, such as a FESS, for power smoothing applications and a high energy density ESS, such as a lithium-ion battery, for reducing the evening peak and avoiding reverse power flow. The main contribution of the proposed method is finding reoccurring daily consumption patterns using the motif discovery algorithm, and using it for the sizing calculations, rather than using an arbitrary day, or reducing the data resolution for the sizing study, as commonly found in literature. Motif discovery can find similar, and yet, not the same patterns using the SAX representation of the data, and generate a representative day for the time series, which can be used for the storage sizing. The detected patterns using motif discovery were used for deriving the characteristic of two types of ESS. Using a simple optimization framework, an optimum cut-off frequency is selected for applying a low-pass filter on the patterns in order to decompose them into low- and high-frequency components, and allocating them to different types of ESS. The nominal capacity, nominal power, and the maximum ramp rate is calculated for each type of ESS using the standard patterns, detected from the measurement data collected at four different low voltage distribution grids in southern Germany with three different data resolutions. It has been shown that the ESS with characteristic derived from only the standard patterns can effectively reduce the evening peak, the reverse power flow, and short-term active power fluctuations for most of the days throughout the whole data set. Moreover, it has been shown, that high-resolution data should be preferred to accurately calculate all the characteristics of an ESS, in particular for power smoothing applications.

In summary, this chapter presented two examples on how techniques and solutions from computer science and information theory can be applied for solving practical problems in power systems. With the increasing number of measurement devices in particular at the low voltage distribution grids, data-driven techniques can become a major tool for providing novel solutions to the challenges in design and operation of low voltage grids.

5 Modelling of High-Speed Flywheel Energy Storage Systems (FESS) for Real-time Simulations

In this chapter, the model of a high-speed Flywheel Energy Storage System (FESS) is presented and simulated in real-time, which considers the losses and auxiliary power requirements of the FESS. The focus of the modelling is on the behavior of the FESS for frequency and voltage support at the low voltage level. The model of the FESS is simulated in real-time with a simulation step of 10 μs, and the results are compared with offline simulation results. The developed real-time simulation model can be used for controller hardware-in-the-loop testing of new controllers, or for the preliminary studies required for Power Hardware-in-the-Loop (PHIL) testing of the FESS, as discussed in Chapter 3. The model of the FESS in this work is based on the structure and design of a specific commercial 60 kW high-speed FESS. However, the model can be easily adjusted for any other systems.

5.1 Potential Applications of Real-time Simulation of a FESS

As discussed in Chapter 3, real-time simulations of new power systems components can accelerate their design, development, and performance verification [96], [100]. A component model, running in real-time, can serve as a *Digital Twin* [99] of that component, which can be used to study its interaction with other hardware. Therefore, engineers have developed models of power systems components including wind turbines [111], [217], photovoltaic systems [218], and superconducting fault current limiters [219], [220] for real-time simulations.

Moreover, a real-time simulation of a power system component together with a power amplifier can be used to build a full-scale emulator of the component. These emulators are often the safer and the cheaper alternatives to the real hardware, for instance, in the case of hazardous lithium-ion batteries or the costly high-speed FESS. Lithium-ion batteries have been emulated at the cell level [221], as small battery energy storage system for household applications [222], and as grid-scale systems for grid-wide applications [223]. Today, commercial and non-commercial emulators are available for many power system components such as wind turbines [143], [224], PV systems [225],

[226], and electrical machines [227], which are used for various applications, including control and converter testing [173], [228], [229].

In addition, as discussed in Chapter 3, real-time simulation of the hardware under test is a recommended step, prior to a PHIL testing. Although offline simulations can also provide useful insights [118], there are advantages of having the model of the hardware under test running in real-time. This includes simulating the hardware model in the same grid model, in which the behavior of the real hardware is later evaluated using PHIL testing.

To our knowledge, the real-time implementation of a FESS model has not yet been carried out before. In this chapter, a model of a high-speed FESS is developed and simulated in real-time, and the results are compared against offline simulation results. In the following chapter, this model is validated using the experimental results obtained from the PHIL testing of a commercial FESS.

5.2 Modeling of a High-speed Flywheel Energy Storage System (FESS)[1]

The model of a FESS comprises the model of each of its individual components. Figure 5.1 illustrates the structure of a high-speed FESS at a component level. The structure of the high-speed FESS presented in Figure 5.1 is based on the design of a 60 kW/3.6 kWh commercial high-speed FESS. The inside view of the container holding the 60 kW FESS and its required auxiliary components is shown in Figure 5.2.

As seen in Figures 5.1 and 5.2 and discussed in Chapter 2, a high-speed FESS consists of a Permanent Magnet Synchronous Machine (PMSM) coupled to a high-inertia rotor, two Voltage Source Converters (VSC) as the Machine-side Converter (MSC) and Grid-side Converter (GSC) with their corresponding controllers, and filters on the AC side of each converter. Due to the extremely low inductance values of the PMSM used in this high-speed FESS, it is necessary to have an LC filter between the PMSM and the machine-side converter to reduce current and torque ripples [230].

[1] Parts of this chapter have been presented in the following publications:
1) S. Karrari, M. Noe, and J. Geisbuesch, "High-speed Flywheel Energy Storage System (FESS) for Voltage and Frequency Support in Low Voltage Distribution Networks," in 3rd IEEE International Conference on Intelligent Energy and Power Systems (IEPS 2018), Kharkov, 2018, pp. 176–182.
2) S. Karrari, M. Noe, and J. Geisbuesch, "Real-time simulation of high-speed Flywheel Energy Storage System (FESS) for distribution networks," in the 9th ACM International Conference on Future Energy Systems (ACM e-Energy 2018), Karlsruhe, 2018.

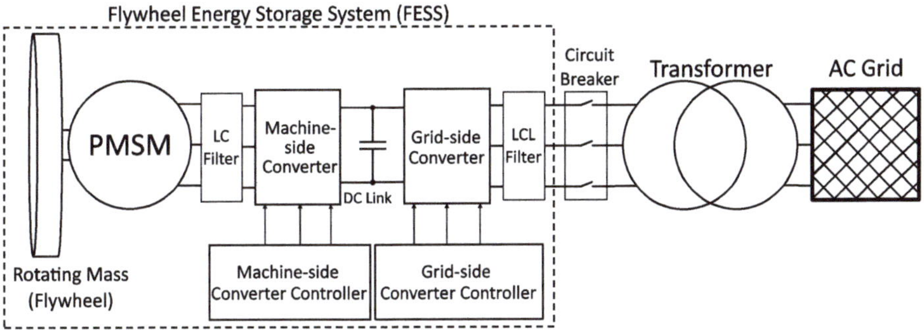

Figure 5.1: Structure of a high-speed Flywheel Energy Storage System (FESS).

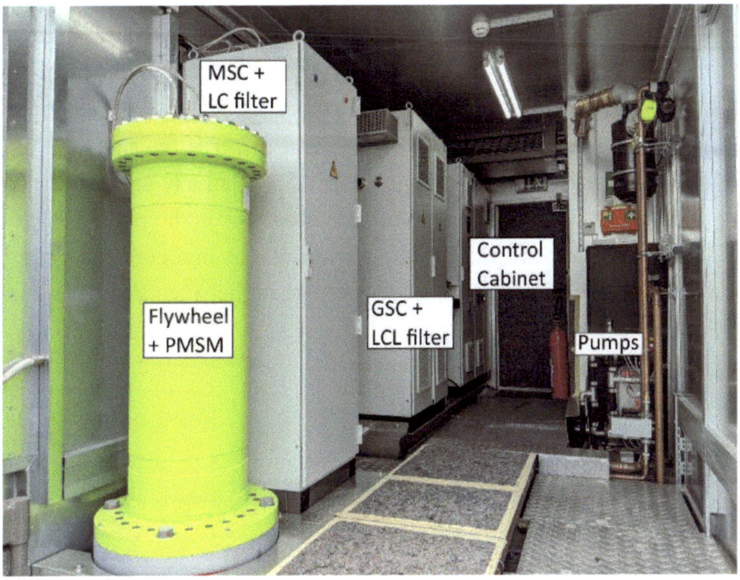

Figure 5.2: The inside view of the container of the 60 kW high-speed FESS.

In principle, the model of a FESS is similar to the model of a variable-speed wind turbine equipped with a PMSM [107]. However, the FESS has faster dynamics, no mechanical input torque, and the ability to absorb energy. In the following section, the model of each individual component of the FESS is discussed in detail.

5.2.1 Modeling of a Permanent Magnet Synchronous Machine

As mentioned in Chapter 2, Permanent Magnet Synchronous Machines (PMSM) are the most common choice for a high-speed FESS, while a low-speed FESS often uses an asynchronous machine. Due to the absence of the field windings, PMSM are more suitable to be used in the vacuum enclosure of high-speed FESS, where convection cooling of the rotor is not possible. They are also simple and robust in their structure, which leads to a higher reliability. Moreover, these machines have high power-to-weight and torque-to-mass ratios and an easier control design, when compared to several other types of electrical machines [73].

Depending on the arrangement of the permanent magnets in the rotor, there are several types of PMSM, with rather diverse characteristics. Four major types of PMSM are shown in Figure 5.3. In a high-speed FESS, a surface mounted PMSM, which is shown in Figure 5.3(a), is often used, in which the permanent magnets are mounted on the rotor surface in a symmetrical arrangement. This leads to a relatively small stator inductance and a negligible reluctance torque. The other common type is the interior PMSM, shown in Figure 5.3(c) and (d), in which the rotor salience results in a non-uniform air gap flux distribution and a high reluctance torque. A comparison between a surface mounted and an interior PMSM is given in Table 5.1. The interior PMSM and the surface mounted PMSM have different characteristics, however, the mathematical description is the same in both cases [231].

Table 5.1: A comparison between the surface mounted and interior PMSM [231].

Type	Advantages	Drawbacks
Surface mounted PMSM	+ Easy construction + Simple design + Uniform air gap + Higher air gap flux density + Lower harmonic content + Simpler control	- Lower mechanical robustness (risk of magnets being detached from the rotor) - More complex speed or position estimation
Interior PMSM	+ Higher mechanical robustness + Higher torque (or acceleration) + Higher overloading capability	- Complex construction - Requires flux barriers

For modeling and description of the notations of a PMSM, a simplified schematic of a two-pole surface mounted PMSM is shown in Figure 5.4. Clearly, the modeling and the equations are also valid for a PMSM with more than two poles.

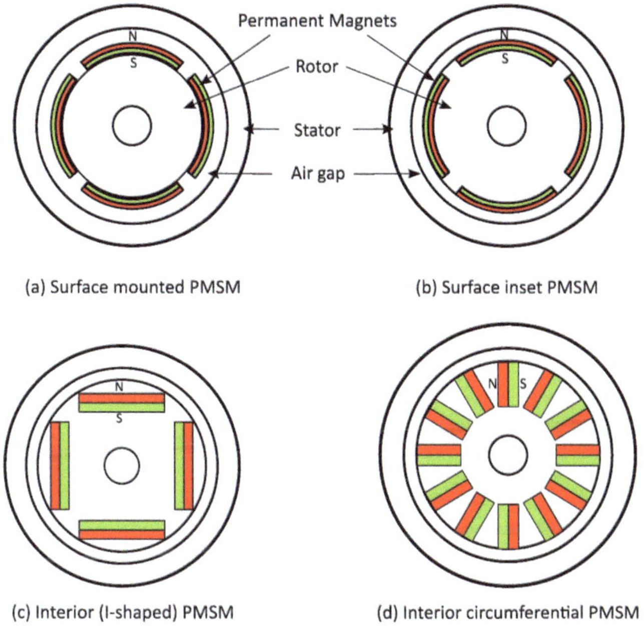

Figure 5.3: Arrangement of the permanent magnets in (a) surface mounted PMSM, (b) surface inset PMSM, (c) interior (I-shaped) PMSM, and (d) interior circumferential PMSM.

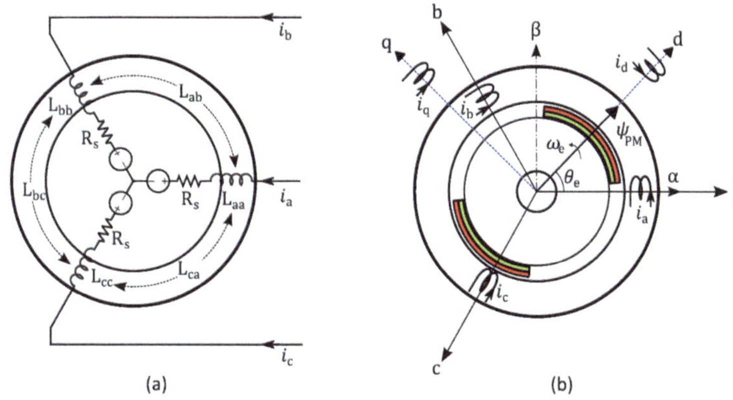

Figure 5.4: (a) The equivalent circuit of a PMSM in the three-phase stationary reference frame. (b) Schematic of a two-pole surface mounted PMSM in the dq-rotating reference frame.

The model of a PMSM has been frequently described in the literature, including in [73], [231]–[233]. In this work, a very detailed model is not necessary, as the goal is to represent the behavior of the FESS at the system level for grid applications. Therefore, the following simplifying assumptions have been made for modeling the PMSM [73]:

- Resistances are constant, i.e., the effect of temperature on the resistances is neglected.
- Capacitances are negligible.
- Hysteresis losses are neglected.
- Eddy current losses are neglected.
- Flux harmonics are neglected.
- The stator windings are identical and positioned in a sinusoidal manner along the air gap. Also, they are connected in a star configuration with an open neutral connection (not grounded), which is common for an inverter connected PMSM.
- As the flux generated by the permanent magnets in the stator is sinusoidal, the magnetic motive force is also sinusoidal.
- The inductance versus rotor position is sinusoidal.

Moreover, the iron saturation is also neglected in this study due to unavailability of the relevant data from the manufacturer. However, for machines with a high power density that are often used, it is recommended to consider the iron saturation [234]. The losses are neglected in the modeling of the PMSM. However, all losses in the FESS including the copper and iron losses in the PMSM are aggregated and considered, as later described in subsection 5.2.5. The permanent magnets are assumed to be a flux linkage source concentrating all their flux along only one axis [231]. Since a surface mounted PMSM is considered, a uniform air gap with no rotor saliency is assumed.

With the aforementioned assumptions, the PMSM can be represented by the equivalent circuit shown in Figure 5.4(a). The relationship between the voltages and currents of each phase for a balanced system in the three-phase stationary reference frame is governed by [73]

$$\begin{bmatrix} u_\mathrm{a} \\ u_\mathrm{b} \\ u_\mathrm{c} \end{bmatrix} = \mathrm{R_s} \begin{bmatrix} i_\mathrm{a} \\ i_\mathrm{b} \\ i_\mathrm{c} \end{bmatrix} + \frac{d}{dt} \begin{bmatrix} \psi_\mathrm{a} \\ \psi_\mathrm{b} \\ \psi_\mathrm{c} \end{bmatrix}, \tag{5.1}$$

in which,

$$\begin{bmatrix} \psi_\mathrm{a} \\ \psi_\mathrm{b} \\ \psi_\mathrm{c} \end{bmatrix} = \begin{bmatrix} \mathrm{L_{aa}} & \mathrm{L_{ab}} & \mathrm{L_{ac}} \\ \mathrm{L_{ba}} & \mathrm{L_{bb}} & \mathrm{L_{bc}} \\ \mathrm{L_{ca}} & \mathrm{L_{cb}} & \mathrm{L_{cc}} \end{bmatrix} \begin{bmatrix} i_\mathrm{a} \\ i_\mathrm{b} \\ i_\mathrm{c} \end{bmatrix} + \psi_\mathrm{PM} \begin{bmatrix} \sin\theta_\mathrm{m} \\ \sin(\theta_\mathrm{m} - \frac{2\pi}{3}) \\ \sin(\theta_\mathrm{m} + \frac{2\pi}{3}) \end{bmatrix}. \tag{5.2}$$

In Eq. (5.1) and (5.2), u_k is the stator phase voltage ($\mathrm{k} \in \{\mathrm{a,b,c}\}$), i_k is the stator phase current, ψ_k is the stator flux linkage, ψ_PM is the permanent magnet flux linkage, $\mathrm{L_{kk}}$ are the self- and mutual inductances of each stator phase, R_s is the stator phase resistance, and θ_m is the mechanical angle of the rotor. According to Eq. (5.1), the voltage of the three-phase windings in a PMSM equals to the sum of the resistive voltage drops and the time derivative of the flux linkages. However, the fluxes and inductances in Eq. (5.2) change nonlinearly with rotation of the rotor. To avoid this, it is common practice to replace the three-phase stationary windings of the stator with two imaginary perpendicular windings, which are rotating with the electrical speed of the rotor. This frame is known as the dq-rotating reference frame and is depicted in Figure 5.4(b). The well-known Park transformation matrix (T_{dq}), as shown in Eq. (5.3), is used for mapping the values from the three-phase stationary frame to the dq-rotating reference frame. In Eq. (5.3), i_d and i_q are the d- and q-axis equivalent stator currents, and i_0, which is the zero component current, is assumed to be zero, as a balanced system is considered. As shown in Figure 5.4(b), it is assumed that the d-axis is aligned with the rotating flux produced by the permanent magnets and the current of the first phase (phase a). The q-axis leads the d-axis by 90 degrees in the counterclockwise direction.

$$\begin{bmatrix} i_d \\ i_q \\ i_0 \end{bmatrix} = \underbrace{\frac{2}{3}\begin{bmatrix} \cos(\omega_e t) & \cos(\omega_e t - \frac{2\pi}{3}) & \cos(\omega_e t + \frac{2\pi}{3})) \\ -\sin(\omega_e t) & -\sin(\omega_e t) & -\sin(\omega_e t + \frac{2\pi}{3}) \\ \frac{1}{2} & \frac{1}{2} & \frac{1}{2} \end{bmatrix}}_{T_{dq}} \begin{bmatrix} i_a \\ i_b \\ i_c \end{bmatrix}. \tag{5.3}$$

Since the reference frame rotates with the same frequency as the supplied voltage in steady-state conditions, this frame is also referred to as the synchronous reference frame in the literature [235]. Using the dq-rotating reference frame, sinusoidal variables become DC, which are much easier to analyze and control. In the dq-rotating reference, the PMSM is modeled as [236]

$$\begin{bmatrix} u_d \\ u_q \end{bmatrix} = R_s \begin{bmatrix} i_d \\ i_q \end{bmatrix} + \frac{d}{dt}\begin{bmatrix} \psi_d \\ \psi_q \end{bmatrix} + \omega_e \begin{bmatrix} -\psi_q \\ \psi_d \end{bmatrix}, \tag{5.4}$$

in which,

$$\begin{bmatrix} \psi_d \\ \psi_q \end{bmatrix} = \begin{bmatrix} L_d i_d + \psi_{PM} \\ L_q i_q \end{bmatrix}, \tag{5.5}$$

and u_d and u_q are the d- and q-axis stator voltage, L_d and L_q are d- and q-axis stator inductance, ψ_d and ψ_q are the stator flux in the d- and q-axis directions, and ω_e is the rotor electrical speed. As seen from Eq. (5.5), using this approach, the flux values are no longer a function of the rotor position. As saturation is neglected, a linear relationship is assumed between the d- and q-axis flux linkages and the stator currents, as shown in

Eq. (5.5). A detailed modeling of the PMSM considering saturation, nonlinear cross coupling, and angular dependencies can be found in [227].

Equations (5.4) and (5.5) can be schematically illustrated by two equivalent circuits, one for each of the d- and q-axis directions, as shown in Figure 5.5. In this figure, the voltage sources represent the induced voltages caused by the rotation of the machine.

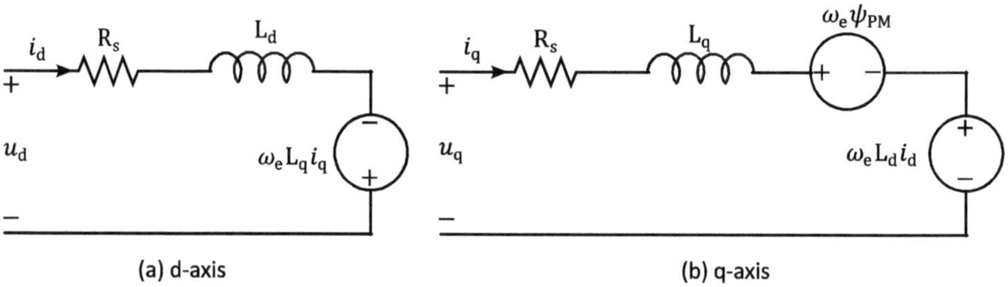

Figure 5.5: Equivalent circuits of a PMSM in the (a) d-axis and the (b) q-axis.

For calculating the speed variations of the rotor, the mechanical part of the PMSM is modeled using the Eq. (5.6)-(5.10), considering the fact that in a FESS no mechanical torque input is applied to the rotor.

$$\omega_e = n_p \omega_m, \tag{5.6}$$

$$\omega_m = \frac{d\theta_m}{dt}, \tag{5.7}$$

$$\tau_e = \frac{3}{2} n_p (\psi_d i_q - \psi_q i_d) = \overbrace{\frac{3}{2} n_p i_d i_q (L_d - L_q)}^{\text{Reluctance torque}} + \overbrace{\frac{3}{2} n_p \psi_{PM} i_q}^{\text{Synchronous torque}}, \tag{5.8}$$

$$J_f \frac{d\omega_m}{dt} = \tau_e - D_f \omega_m, \tag{5.9}$$

$$E_{FW} = \frac{1}{2} J_f \omega_m^2. \tag{5.10}$$

In these equations, ω_m is the mechanical angular velocity of the rotor, τ_e is the electrical torque, n_p is the number of pole pairs, D_f is the friction coefficient, E_{FW} is the energy stored in the rotor, and J_f is the combined moment inertia of the PMSM and the flywheel. Equation (5.9) determines the variations in the speed of the rotor or flywheel, from which the changes in its energy content can be calculated according to Eq. (5.10). The term $D_f \omega_m$ in Eq. (5.9) is referred to as the drag toque, which is minimized in a high-speed FESS with the use of a vacuum enclosure and magnetic bearings. Equation (5.10) resembles the swing equation of conventional synchronous generators. This similarity is used later in Chapter 7 to introduce an inertia emulation controller for a FESS in low-inertia grids.

The electrical torque in Eq. (5.8), consists of two parts; the reluctance torque and the synchronous torque. In a surface mounted PMSM, the d- and q-axis stator inductances (L_d and L_q) are almost equal, making the reluctance torque nearly zero. Therefore, Eq. (5.11) can be further simplified to

$$\tau_e = \frac{3}{2} n_p \psi_{PM} i_q. \tag{5.11}$$

Since the permanent magnet flux is assumed to be constant[2], it can be seen from Eq. (5.11) that the electromagnetic torque can be controlled through the q-axis component of the stator current. This simplicity cannot be achieved when using other reference frames such as the stationary αβ-frame.

The active (P_e) and reactive power (Q) of a PMSM in the dq-rotating reference frame is calculated using Eq. (5.12) and (5.13), which is also known as the instantaneous power theory [237].

$$P_e = \frac{3}{2}\left(u_d i_d + u_q i_q\right), \tag{5.12}$$

$$Q = \frac{3}{2}\left(-u_d i_q + u_q i_d\right). \tag{5.13}$$

The model of the PMSM in the dq-rotating reference frame can also be rewritten as a nonlinear state-space model by defining i_d, i_q, and ω_e as the state variables, and u_d and u_d as the system input, which result

$$\frac{d}{dt}\begin{bmatrix} i_d \\ i_q \\ \omega_e \end{bmatrix} = \overbrace{\begin{bmatrix} \frac{-R_s}{L_d} & 0 & 0 \\ 0 & \frac{-R_s}{L_q} & \frac{-\psi_{PM}}{L_q} \\ 0 & \frac{3}{2}\frac{n_p^2}{J_f}\psi_{PM} & \frac{-D}{J_f} \end{bmatrix}\begin{bmatrix} i_d \\ i_q \\ \omega_e \end{bmatrix} + \begin{bmatrix} \frac{1}{L_d} \\ \frac{1}{L_q} \\ \frac{-n_p}{J_f} \end{bmatrix}\begin{bmatrix} u_d \\ u_q \\ 0 \end{bmatrix}}^{\text{Linear part}} + \overbrace{\begin{bmatrix} \frac{L_q}{L_d}\omega_e i_q \\ -\frac{L_q}{L_d}\omega_e i_d \\ \frac{3}{2}\frac{n_p^2}{J_f} i_d i_q (L_d - L_q) \end{bmatrix}}^{\text{Nonlinear part}}. \tag{5.14}$$

Knowing the exact values of the machine parameters is mandatory for an accurate simulation of its dynamic and steady-state behavior. The permanent magnet flux (ψ_{PM})

2 Minor variations due to temperature changes may be present in practice, which are neglected in this work.

can be easily calculated using an open-circuit test with a reasonable accuracy. However, determining the precise values for d- and q-axis stator inductances (L_d and L_q) can be more complex, as saturation and cross coupling effects are also affecting the results. Various analytical, finite element, and experimental methods can be used to estimate these values, as described fully in [238], [239]. In this work, these values are provided by the manufacturer and are given in Table 5.2.

Table 5.2: Parameters of the PMSM of the high-speed FESS according to the manufacturer.

Parameter	**Symbol**	**Value**
Stator resistance	R_s	9 mΩ
d- and q-axis stator inductance	L_d, L_q	415.8 μH
Permanent magnet flux linkage	ψ_{PM}	0.1178 Vs
Number of pole pairs	n_p	1
Rotor (flywheel) inertia	J_f	1.26 kgm^2
Friction coefficient	D_f	4.0898×10^{-5} Nms
Nominal electrical power	P_n	60 kW
Maximum electrical power (short-term)	P_{max}	91.5 kW
Maximum stator current	i_{smax}	141 A
Maximum electrical torque	τ_{emax}	21 Nm
Maximum mechanical angular velocity	ω_{max}	4712.4 rad/s

5.2.2 Modeling of Voltage Source Converters

As shown in Figure 5.1, a FESS is controlled via two voltage source converters in a back-to-back configuration. As discussed in Chapter 2, several converter topologies can be used in a FESS, but the two-level and three-level converters are the most common topologies found in commercial systems. In the high-speed FESS in this work, both converters have a two-level topology.

Depending on the intended study, there are various well-known converters models that can be selected, which differ mainly on how detailed the switches and semiconductors are represented. Comparative studies of different converter models for various applications can be found in [110], [240], [241]. Clearly, more detailed models have a higher accuracy, but at the cost of a higher computational effort. A small computation time is of crucial importance for real-time simulations, in which all calculations are required to be completed within a short simulation time step. Therefore, choosing the appropriate level of details is necessary to limit the model computation time, while simultaneously having an acceptable model accuracy. Average model, switching model, and detailed models are among the most commonly used converter models. These well-known models are explained briefly in this section, followed by a discussion on choosing the right model for the real-time simulations in this work. It should be noted that there are several other less-known converter modeling techniques, such as the sub-cycle average models [242], which are not discussed here.

5.2.2.1 Average Model

The average model is based on the assumption that the switching frequency of the converter is much greater than the fundamental frequency of the grid. In this model, the switching transitions, harmonics, and ripples are removed from the voltages and currents by averaging over one switching period. Nevertheless, lower frequency components of voltages and currents are preserved.

In the average model, all switches are assumed to be operated in the continuous conduction mode [235]. The semiconductors like IGBTs and diodes are not explicitly modeled, and the converter on the AC side is simply modeled as a three-phase controlled AC voltage source with a small internal resistance using [243]

$$u_{\mathrm{cj}} = m_{\mathrm{j}} \frac{U_{\mathrm{DC}}}{2} - r_{\mathrm{on}} i_{\mathrm{cj}}. \qquad (5.15)$$

In Eq. (5.15), m_{j} is the modulation index of each phase ($\mathrm{j} \in \{\mathrm{a,b,c}\}$), which is determined by the converter controller, U_{DC} is the voltage of the DC side, u_{cj} and i_{cj} are the phase voltage and current of the AC side, respectively, and r_{on} represents the on-state resistance of the switches [243]. It is important to ensure that $|m_{\mathrm{j}}| \leq 1$ at all times in order to guarantee a linear Pulsed Width Modulation (PWM). Equation (5.15) can be rewritten in the dq-rotating reference frame, rotating at the fundamental frequency of the AC side voltage using [243]

$$\begin{bmatrix} u_{\mathrm{cd}} \\ u_{\mathrm{cq}} \end{bmatrix} = \frac{U_{\mathrm{DC}}}{2} \begin{bmatrix} m_{\mathrm{d}} \\ m_{\mathrm{q}} \end{bmatrix} - r_{\mathrm{on}} \begin{bmatrix} i_{\mathrm{cd}} \\ i_{\mathrm{cq}} \end{bmatrix}. \qquad (5.16)$$

In Eq. (5.16), u_{cd} and u_{cq} are the d- and q-axis voltage of the AC side of the converter, i_{cd} and i_{cq} are the d- and q-axis current, and m_{d} and m_{q} are the d- and q-axis component of the modulation index, respectively.

For modeling the DC side, a lossless converter is assumed (losses are later aggregated in subsection 5.2.5). Therefore, the current of the DC side of the converter is calculated using

$$I_{\mathrm{DC}} = \frac{P_{\mathrm{AC}}}{U_{\mathrm{DC}}} = \frac{\frac{3}{2}\left(u_{\mathrm{cd}} i_{\mathrm{cd}} + u_{\mathrm{cq}} i_{\mathrm{cq}}\right)}{U_{\mathrm{DC}}}. \qquad (5.17)$$

According to Eq. (5.17), the DC side of the converter is modeled as a controlled DC current source, in which the current is a function of the active power flowing through the AC side (P_{AC}). An equivalent circuit of the average model of a voltage source converter is shown in Figure 5.6.

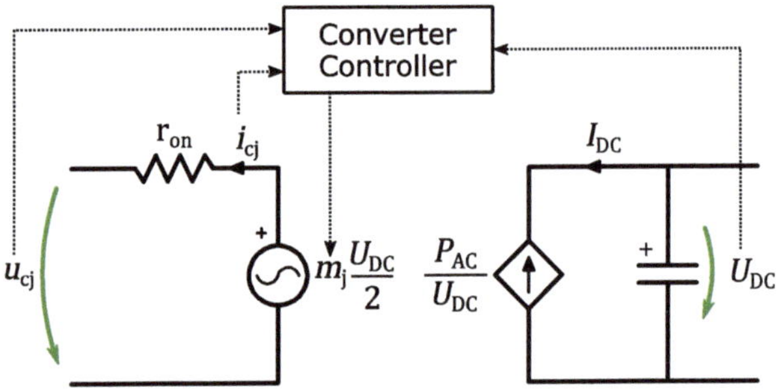

Figure 5.6: The equivalent circuit of the average model of a voltage source converter.

Since all switching transitions, harmonics, and ripples are neglected, the average model requires a very low computational effort. It is the most common model used for designing converter controllers, as it provides a simple physical insight. Controllers can be included in full detail, considering the inner, e.g., current, and outer, e.g., power or DC-link voltage, control loops. From the grid perspective, since the high-frequency harmonics are mostly attenuated by filters, the average model can be suitable for integration studies of distributed energy resources into power systems. It can be used for transient stability studies of fast phenomena in the order of tens to hundreds of milliseconds [110]. However, this model is not suitable for the investigation of harmonics or internal converter analyses.

5.2.2.2 Switching Model

In the simplest form of switching model, the switches are modeled as ideal on/off switches. The switching transitions such as dead times are ignored based on the assumption that they take only a small fraction of the switching period. Each switch is controlled by a switching function, which has the value of one, when the switch is on, and zero, when the switch is off. On-state losses can be added as a resistor in series with the ideal switches. Therefore, the voltage of each converter phase using the switched model is calculated using [243]

$$u_{cj} = (S_{mj} - S_{nj})\frac{U_{DC}}{2} - r_{on} i_{cj}, \tag{5.18}$$

where S_{mj} and S_{nj} are the switching function of two complementary switches for each converter leg, and

$$S_{\mathrm{mj}} + S_{\mathrm{nj}} = 1 \tag{5.19}$$

at all times, in order to avoid short-circuiting of the DC-link. The switching functions are the gate signals from the pulsed width modulation controller. Modeling of the DC side is similar to the one from the average model by calculating the active power from the three phases voltages and current. The equivalent circuit for a voltage source converter using the switching model is shown in Figure 5.7

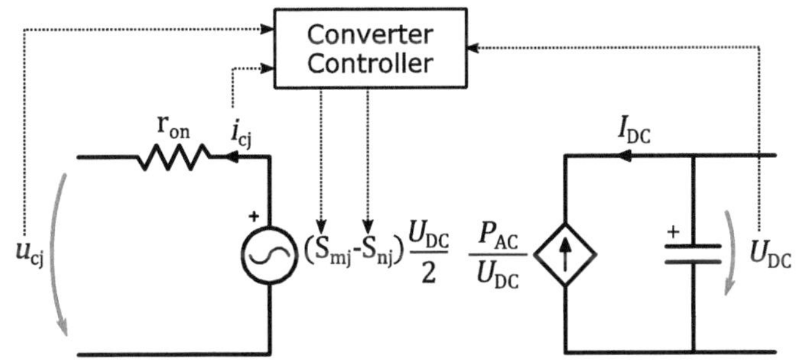

Figure 5.7: The equivalent circuit of the switching model of a voltage source converter.

In a more detailed approach, switches can be modeled as two-value resistors, having a high resistance when they are in the off-state (in the range MΩ) and a low resistance, when they are in the on-state (in the order of mΩ). The snubber circuit and the anti-parallel diodes are also modeled in the same manner. The choice between high and low resistance values for the switches is made on the basis of the gate signals received from the modulation controller, but this choice for the anti-parallel diodes are made based on the instantaneous voltage and current values.

Switching models can represent voltage and current harmonics, which is a significant advantage compared to the average model. However, small simulation time steps are required usually for simulating such models to avoid numerical instability, in particular for high switching frequencies.

5.2.2.3 Detailed Model

There are different types of detailed converter models in literature, as described in [240]. In one approach, switches are represented by variable resistors, which can either follow a simplified linear function or a nonlinear one according to the voltage-current characteristics of the semiconductors, given in the manufacturer's datasheets [107]. The detailed modeling of converters can be even more complex, in which all the parasitic capacitors of the switches, stray inductances, and the packaging effects are considered.

Simulating such models usually requires extremely small simulation time steps to accurately represent fast switching events and solve the stiff differential equations of the system with high order solvers. The combination of small simulation time steps and high model complexity leads to a long run time for these models. Therefore, such models are often used for a specific part of the system and for a short time period of a few switching cycles. They are mostly used for specific applications, such as simulating internal converter faults, studying switching transitions, switching losses and electromagnetic interference (EMI). This level of details is not required for power system integration studies or designing and testing control systems [110], [240], [244], which is the intention in this work.

5.2.2.4 Selecting the Right Converter Model for Real-time Simulations

As discussed in Chapter 3, in real-time simulations, equations should be solved within a fixed simulation time step. The average model can easily be simulated in real-time, independent of the switching frequency, as variations within each switching period are neglected. Real-time simulation of a switching model for converters can be challenging for high switching frequencies. As discussed in subsection 3.1.2, in practice, the simulation sampling frequency is usually chosen in the range of 50 to 100 times of the switching frequency for an accurate simulation of converter switching models [111], [245]. Furthermore, in the switching model using two-value resistors, there is a sudden change between a high-value resistance and a low-value one. Hence, an even smaller time step may be required to avoid numerical instabilities, when using the fixed-step non-iterative solvers of the real-time simulators [107]. Therefore, real-time simulation of converters with a high PWM carrier frequency using the switching models may require simulation time steps of less than few microseconds, which is not possible using CPU-based real-time simulators. In this case, the model has to be simulated on an FPGA with a submicrosecond simulation time step.

In this work, two voltage source converters with a PWM carrier frequency of 8 kHz have to be simulated in real-time. This requires a simulation step of 1.25-2.5 μs for accurate real-time simulation of the switching models according to [111], [245], which is not possible using CPU-based real-time simulators. However, as suggested in [110], for applications such as frequency and voltage control in power systems, which are the applications intended in this work, the use of average converter models can be adequate. Therefore, average converter models using Eq. (5.15)-(5.18) are used for the modeling of the FESS. This also allows using larger simulation time steps, which means more components can be simulated in real-time in a single model. It should be noted that simulating the switching model of these converters on the FPGA or with the help of switching event compensation algorithms [246] is also possible, but deemed unnecessary for this application.

5.2.3 Grid-side Converter Controller

Figure 5.8 depicts the block diagrams of the grid-side converter and machine-side converter controllers of the high-speed FESS, which are designed for the applications of voltage and frequency support in low voltage grids. Description of the notations and values of parameters for the grid-side and machine-side converter controllers are provided in Table 5.3 and Table 5.4, respectively.

The grid-side converter controller has two major tasks:

- Supporting the grid voltage according to the latest German grid code for the low voltage grids [182], by controlling the reactive power exchange between the FESS and the grid.
- Maintaining the DC-link voltage on the reference value, by controlling the active power exchanged between the DC-link capacitor and the grid.

Grid voltage support: Since the FESS in this work is connected to the low voltage distribution grid, it has to comply with the latest grid connection codes for this voltage level. Since November 2018, the new version of the grid code VDE-AR-N 4105 [182] is applicable for the German grids, which indicates that any ESS larger than 4.6 kVA, should follow the Q(U) characteristics, depicted in Figure 4.5(b). The Q(U) characteristics requires the ESS to start absorbing or injecting reactive power when the voltage passes the ±3 % of the nominal value and reaches its maximum reactive power capability when the voltage reaches ±7 %. The gradient of the Q(U) characteristics determines the voltage droop coefficient D_q, which indicates how much reactive power should be injected or absorbed with respect to the voltage changes. The voltage deadband u_{db} is set to ±3 %, which helps avoiding continuous operation of the FESS during minor voltage variations. The Q(U) characteristic replaces the grid code requirement from 2011, which indicated that reactive power support is only required if the ESS injects power larger than 50 % of its rated value to the grid.

In this work, the instantaneous grid voltage (u_g) is calculated in the dq-rotating reference frame and is used for the calculation of the reactive power reference, which enables an immediate detection of any changes in the voltage. This is calculated using

$$u_{g} = \sqrt{u_{dg}^{2} + u_{qg}^{2}}. \tag{5.20}$$

The reactive power reference for grid voltage support is generated according to the value of u_g, the voltage deadband u_{db}, and the voltage droop coefficient D_q. The reference value for the reactive power can also sent by an operator using Q_{cref} (see Figure 5.8).

DC-link voltage control: A conventional PI controller with saturation is used for controlling the DC-link voltage. The output of the PI controller determines the active power reference for the grid-side converter controller (P_{gref}), which is limited to the maximum power of the FESS (P_{max}).

Determining the current reference values: The upper level controllers or commands by the operator determine the active and reactive power reference values (P_{gref} and Q_{gref}). The instantaneous power theory [237] is used for calculating the d- and q-axis current reference values (i_{qgref} and i_{dgref}) from the active and reactive power set points. Similar to the PMSM, the active and reactive power of the grid-side converter can be calculated from the voltages and currents in the dq-rotating reference frame using

$$P_{g} = \frac{3}{2}\left(u_{dg} i_{dg} + u_{qg} i_{qg}\right), \tag{5.21}$$

and

$$Q_{g} = \frac{3}{2}\left(-u_{dg} i_{qg} + u_{qg} i_{dg}\right). \tag{5.22}$$

The dq-rotating reference frame of the grid-side converter controller is synchronized by the phase angle of the grid voltage (θ_{PLL}), which is measured by a conventional Synchronous Reference Frame Phase-locked Loop (SFR-PLL). It is assumed that the dq-rotating reference frame is chosen in a way, that the grid voltage is fully aligned with the d-axis, resulting in a q-axis component of zero for the voltage, i.e., $u_{qg} = 0$. The phase-locked loop controls the rotating reference frame angle, in order to maintain this condition with the changes in the system operating points (see subsection 5.2.5). By doing so, according to Eq. (5.21) and (5.22), the active and reactive power can then be independently controlled by controlling the values of i_{dg} and i_{qg}, respectively. Therefore, the d- and q-axis reference currents are calculated using

$$i_{dgref} = \frac{2P_{gref}}{3u_{dg}}, \tag{5.23}$$

and

$$i_{\mathrm{qgref}} = \frac{-2Q_{\mathrm{gref}}}{3u_{\mathrm{dg}}}. \tag{5.24}$$

The reference values for the d- and q-axis currents are sent to the inner current controller. The inner current controllers are conventional PI controllers. As seen in Figure 5.8, the outputs of the PI current controllers are added with decoupling terms in order to enable an independent control of the d- and q-axis currents. A feedforward term has also been added to reduce the high transient currents during the converter start-up, which also decouples dynamics of the converter system from those of the grid, and improves its disturbance rejection capability, as suggested in [243]. At the end, the grid-side converter controller generates the modulation index m_{dg} and m_{qg} in the dq-rotating reference frame, which are transformed back to the stationary frame using the inverse Park transform and the grid phase angle measured by the phased-locked loop.

Table 5.3: Variables definition and parameters of the grid-side converter (GSC) controller.

Parameter	**Symbol**	**Value**
Instantaneous grid voltage	u_{g}	-
Grid voltage reference	u_{ref}	325.27 V
Voltage deadband	$\mathrm{u_{db}}$	9.76 V (3%)
Voltage droop coefficient	Dq	6.52 kVar/V
External reactive power command	$\mathrm{Q_{crefg}}$	-
Total reactive power reference of the GSC	$\mathrm{Q_{gref}}$	-
Maximum reactive power of the GSC	$\mathrm{Q_{max}}$	34 kVar
d-axis and q-axis component of the grid voltage	$u_{\mathrm{dg}}, u_{\mathrm{qg}}$	-
d-axis and q-axis current reference of the GSC	$i_{\mathrm{dgref}}, i_{\mathrm{qgref}}$	-
d-axis and a-axis grid current of the GSC	$i_{\mathrm{qg}}, i_{\mathrm{dg}}$	-
Three-phase grid current	$i_{\mathrm{ag}}, i_{\mathrm{bg}}, i_{\mathrm{cg}}$	-
Active power reference of the GSC	P_{gref}	-
DC-link capacitor voltage	U_{DC}	-
DC-link capacitor voltage reference	U_{DCref}	720 V
q-axis current controller of the GSC- proportional term	$\mathrm{K_{P1g}}$	5
q-axis current controller of the GSC- integral term	$\mathrm{K_{I1g}}$	100
DC-link voltage controller - proportional term	$\mathrm{K_{P2g}}$	20
DC-link voltage controller - integral term	$\mathrm{K_{I2g}}$	1000
d-axis current controller of the GSC- proportional term	$\mathrm{K_{P3g}}$	5
d-axis current controller of the GSC- integral term	$\mathrm{K_{I3g}}$	100
Estimated phase angle from the phase-locked loop	θ_{PLL}	-
d-axis and q-axis modulation index of the GSC	$m_{\mathrm{dg}}, m_{\mathrm{qg}}$	-
Three-phase modulation index of the GSC	$m_{\mathrm{ag}}, m_{\mathrm{bg}}, m_{\mathrm{cg}}$	-

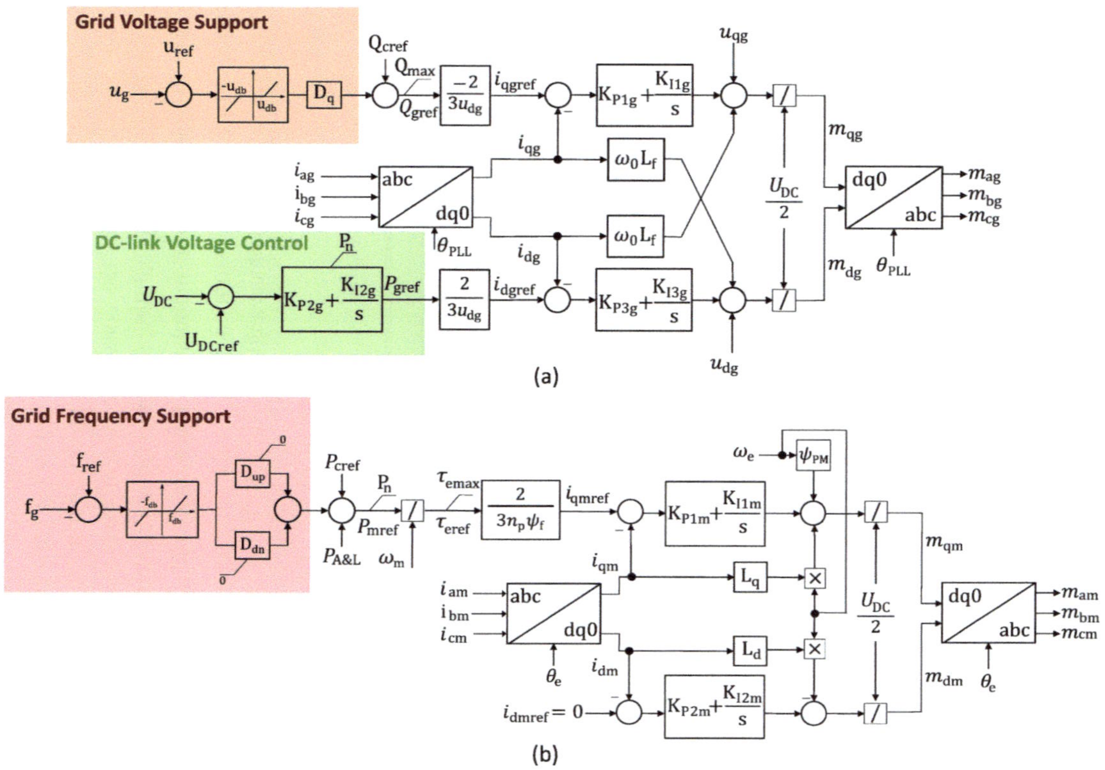

Figure 5.8: (a) Block diagram of the Grid-side Converter (GSC) controller. The upper part of the controller is responsible for supporting the grid voltage, while the lower part controls the DC-link voltage. (b) Block diagram of the Machine-Side Converter (MSC) controller. The upper part is responsible for the providing frequency support to the grid. The Maximum Torque per Ampere (MTPA) control is implemented for controlling the PMSM.

5.2.4 Machine-side Converter Controller

The machine-side converter controller controls the active power flow between the PMSM and the DC-link capacitor through controlling the currents of the machine. The active power reference can either be set by the grid frequency support block or by the operator through P_{cref}. The losses and auxiliary power required for running the entire system, represented by $P_{L\&A}$ in Figure 5.8(b), should also be considered for providing a certain power at the grid connection point. This is due to the fact, that this particular FESS has no additional auxiliary power connection, and all auxiliary components are also powered from its main terminal.

The active power of the flywheel is transferred to the grid by the grid-side converter controller, as it attempts to maintain the DC-link voltage. A description of the notations and values of parameters of the machine-side converter is provided in Table 5.4.

Grid frequency support: Similar to the voltage support requirements, the FESS has to follow the frequency support requirement of the latest German grid code for low voltage grids, the VDE-AR-N 4105:2018-11 [182]. This grid code indicates that the FESS has to stay connected to the grid in the frequency range of 47.5 Hz to 51.5 Hz, which is also known as the frequency ride-through. In this range, the FESS has to follow the P(f) characteristic shown in Figure 4.5(a). According to the P(f) characteristic the frequency deadband (f_{db}) for systems connected to the low voltage grids is 200 mHz. Also, the grid code requires different frequency droop coefficients for over- and under-frequency events larger than the deadband. Hence, the droop for the over-frequency (D_{up}) and under-frequency events (D_{dn}) are separated in Figure 5.8. In case of an over-frequency disturbance in the range of 50.2 to 51.5 Hz, the FESS has to participate in the frequency regulation by absorbing power with the rate of 40 % per Hz, after deducting the frequency deadband. In case of under-frequency events, the FESS has to inject power to the grid with the rate of 100 % per Hz until reaching its maximum power. Therefore, in Figure 5.8(b),

$$P_{mref} = \begin{cases} 20P_n \dfrac{50.2 - f_g}{50\ \text{Hz}} & \text{for } 50.2\ \text{Hz} \leq f_g \leq 51.5\ \text{Hz} \\ 50P_n \dfrac{49.8 - f_g}{50\ \text{Hz}} & \text{for } 47.5\ \text{Hz} \leq f_g \leq 49.8\ \text{Hz} \end{cases}, \tag{5.25}$$

where P_n is the nominal power of the FESS and f_g is the grid frequency.

Speed-dependent ramp-rate limiter: In practice, a FESS can only deliver the nominal power above a particular minimum rotational speed. This is due to the fact, that at low rotational speeds, providing the nominal power requires high amounts of torque, which can be above the maximum permissible torque of the PMSM. Therefore, it is important to consider the torque limiter in modeling a FESS, as shown in Figure 5.8(b). For this particular FESS, the torque limiter limits the power ramp rate to approximately 410 W/s, until reaching the speed of 23,200 rpm.

Determining the current reference values: There are several methods for controlling a PMSM [231], [235], [244], [247]. A categorization of some of the most well-known methods is shown in Figure 5.9. Selecting the most appropriate control method depends on the structure and type of the PMSM, and its main application. Conventional Field-Oriented Control (FOC) methods are generally more preferred due to a simpler design.

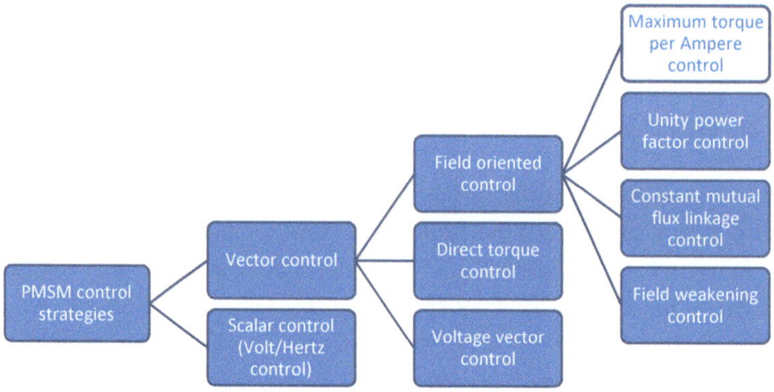

Figure 5.9: Different control strategies for controlling a PMSM (figure adopted from [247]).

For the surface mounted PMSM used in the high-speed FESS, the Maximum Toque per Ampere (MTPA) control has been used. This method is suitable for applications in which fast reaction times are required. With the use of the maximum torque per ampere control, the copper losses of the machine is also minimized [248]. As shown in Eq. (5.8), in a surface mounted PMSM, the electrical torque is a function of the q-axis component of the stator current only. Therefore, for a given stator current, the maximum torque is achieved, when the current is fully aligned with the q-axis. Therefore, to implement the maximum torque per ampere control, the d-axis component of the current should be forced to zero, i.e., $i_{\mathrm{dmref}} = 0$. Figure 5.10(a) shows the vector diagram of the stator current without the maximum torque per ampere control, while Figure 5.10(b) depicts the case when the maximum torque per ampere is applied. This method is also known as the constant torque angle control, since the torque angle, δ in Figure 5.10, is always kept at 90 degrees with respect to the flux linkage of the permanent magnets (ψ_{PM}). The reference value for the q-axis current (i_{qmref}) is calculated from Eq. (5.8) from the reference value for the electrical torque (τ_{eref}) using

$$i_{\mathrm{qmref}} = \frac{2\tau_{\mathrm{eref}}}{3n_{\mathrm{p}}\psi_{\mathrm{PM}}}. \tag{5.26}$$

The torque reference itself is calculated from the active power reference of the machine-side converter controller (P_{mref}) and the instantaneous speed of the FESS (ω_{m}). Conventional PI controllers with decoupling terms have been used for the inner current controllers, similar to the grid-side converter.

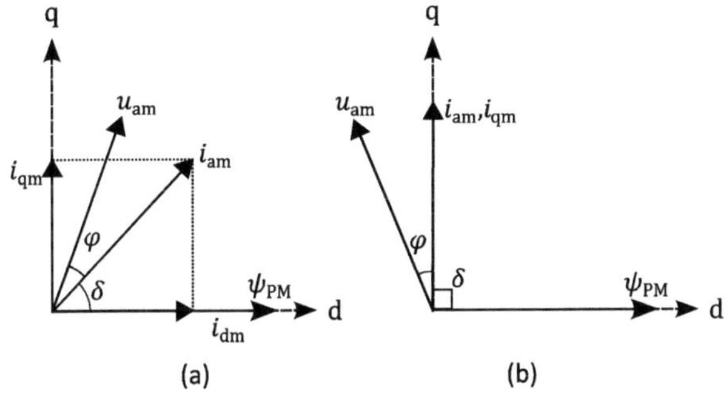

Figure 5.10: Vector diagram of the current and the voltage of the PMSM (a) without the maximum torque per ampere control, and (b) with the maximum torque per ampere control.

Table 5.4: Variables definition and parameters of the machine-side converter (MSC) controller.

Parameter	**Symbol**	**Value**
Grid frequency	f_{g}	-
Grid frequency reference	$\mathrm{f_{ref}}$	50 Hz
Frequency deadband	$\mathrm{f_{db}}$	200 mHz
Over-frequency frequency droop coefficient	$\mathrm{D_{up}}$	24 kW/Hz
Under-frequency frequency droop coefficient	$\mathrm{D_{dn}}$	60 kW/Hz
External active power command	P_{crefm}	-
Total active power reference of the machine-side converter	P_{mref}	-
Losses and auxiliary power	$P_{\mathrm{L\&A}}$	-
Electrical torque reference	τ_{eref}	-
q-axis current controller of the MSC- proportional term	$\mathrm{K_{P1m}}$	1
q-axis current controller of the MSC- integral term	$\mathrm{K_{I1m}}$	3000
d-axis current controller of the MSC- proportional term	$\mathrm{K_{P2m}}$	1
d-axis current controller of the MSC - integral term	$\mathrm{K_{I2m}}$	3000
d-axis and q-axis current reference values of the MSC	$i_{\mathrm{dmref}}, i_{\mathrm{qmref}}$	-
d-axis and q-axis currents of the MSC	$i_{\mathrm{dm}}, i_{\mathrm{qm}}$	-
Three-phase currents of the MSC	$i_{\mathrm{am}}, i_{\mathrm{bm}}, i_{\mathrm{cm}}$	-
d-axis and q-axis modulation index	$m_{\mathrm{dm}}, m_{\mathrm{qm}}$	-
Three-phase modulation index	$m_{\mathrm{am}}, m_{\mathrm{bm}}, m_{\mathrm{cm}}$	-

It is important to note that in high-speed machines, such as the PMSM in a high-speed FESS, the speed and position of the rotor are often not directly measured, and are estimated from the voltage and current values, with the methods described in [249], [250]. However, since these estimation methods often show a good accuracy at high speeds [249], in which the FESS is often operated, the estimation procedure is neglected in this work, and rotor speed is directly calculated from the machine model. This is also necessary to limit the model computation time for real-time simulations.

5.2.5 Modeling of Synchronous Reference Frame Phase-Locked Loop

For some applications, including frequency support in low voltage and low inertia grids (see Chapter 7), the dynamics of the phased-locked loop should also be modelled [110]. In this chapter, a Synchronous Reference Frame Phase-locked Loop (SRF-PLL) with a low-pass filter, as shown in Figure 5.11, is used for tracking the frequency and phase of the grid voltage. The estimated phase angle by the phase-locked loop (θ_{PLL}) is used in the grid-side converter controller to synchronize the dq-rotating reference frame with the grid voltage. This is also necessary for connecting the FESS to the grid with minimum transient currents.

As seen in Figure 5.11, the grid voltage, after being converted to its per unit value, is multiplied by the cosine of the phase estimated by the phased-locked loop. By assuming that the grid voltage in per unit is $\sin(\omega t+\varphi)$, this results in

$$\sin(\omega t+\varphi)\cos(\theta_{\mathrm{PLL}})=\frac{1}{2}\left(\underbrace{\sin(\omega t+\varphi+\theta_{\mathrm{PLL}})}_{\text{Periodic component at }2\omega}+\underbrace{\sin(\omega t+\varphi-\theta_{\mathrm{PLL}})}_{\text{DC component}}\right). \tag{5.27}$$

If the phase estimated by the phase-locked loop is approximately close to the phase of the input signal, i.e., $\theta_{\mathrm{PLL}}\approx\omega t+\varphi$, then Eq. (5.27) consists of a term with double the fundamental frequency and a DC component close to zero. This is due to the fact, that $\sin(\omega t+\varphi-\theta_{\mathrm{PLL}})$ in case of small phase angles is approximated by $\omega t+\varphi-\theta_{\mathrm{PLL}}$. The moving average block in Figure 5.11 eliminates the periodic component by averaging at the fundamental frequency. Therefore, only the DC component remains, which reflects the error in estimating the phase of the input signal by the phase-locked loop. A PI controller forces the steady-state error to zero. The output of the PI controller is the electrical angular velocity. A simple numerical integration generates the phase output of the phase-locked loop. The angular velocity is then converted to Hertz and passed through a low-pass filter and rate limiter to reduce the effect of harmonics and asymmetries of the input voltage on the estimated frequency [251]. The PI controller parameters used for the phase-locked loop in this work are tuned to have a quick response to step changes in the frequency, and are given in Table 5.5.

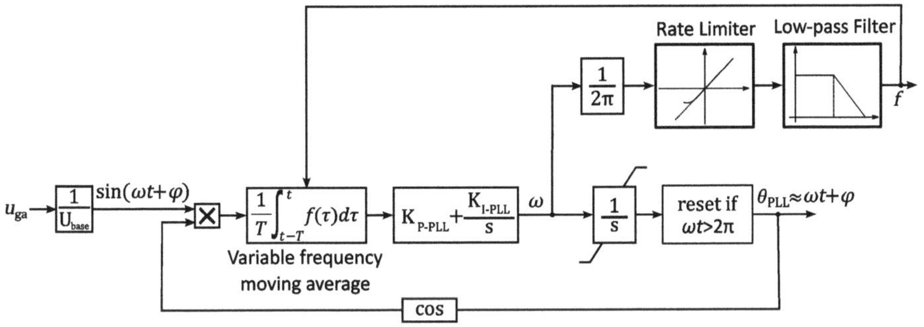

Figure 5.11: Block diagram of a synchronous reference frame phase-locked loop with a low-pass filter.

Table 5.5: Variables definition and parameters of the synchronous reference frame phase-locked loop (SRF-PLL).

Parameter	Symbol	Value
SRF-PLL PI controller- proportional term	K_{P-PLL}	90
SRF-PLL PI controller - integral term	K_{I-PLL}	10

5.2.6 Modeling of System Losses and Auxiliary Power Requirements

For obtaining an accurate model of a FESS, it is also necessary to model the losses within the system. The losses in a FESS can be categorized in three main groups, which are:

- **Speed-dependent losses:** These are the losses that scale with the angular velocity of the flywheel (ω_m). Despite the fact that the high-speed FESS operates in a vacuum enclosure, the vacuum is not an absolute vacuum, with a pressure of just below 0.03 mbar during normal operation. Therefore, the air drag still exists, although being significantly small. The losses associated with air drag is proportional to the square of the rotational speed [252]. In addition, the eddy current and hysteresis losses in the PMSM also scale with the angular velocity of the rotor [66], [253], which can be added to the drag losses.
- **Current-dependent losses**: These are additional losses which appear only during the time when the FESS is being charged or discharged, and scale with the square of the current drawn from or injected to the FESS (i_s). This includes

the copper losses in the PMSM, the two voltage source converters, filters, and other conductors within the FESS.

- **Constant losses:** This includes any losses within the system components that are constant throughout the whole operation range of the FESS.

As discussed, the high-speed FESS in this work has only one grid connection point. Therefore, the power required for all the axillary components including the cooling water and vacuum pumps, active magnetic bearings, control and measurement systems, fans and container ventilation system, fire and safety systems, and all other components are also provided by the flywheel. The auxiliary power is constant for some components, such as the power supply for the controllers, while for others, such as for the cooling systems, can vary depending on the system operating conditions, including the speed and the current of the FESS.

Due to the complex structure of the high-speed FESS with numerous components, the losses and auxiliary power are aggregated all together and estimated directly from measurement, rather than using an analytical approach to calculate each part individually. The three types of losses in the FESS and the axillary power required for running the system are aggregated and modeled in one variable, $P_{\mathrm{L\&A}}$, which is calculated using

$$P_{\mathrm{L\&A}} = c_1 + c_2 i_{\mathrm{s}}^2 + c_3 \omega_{\mathrm{m}}^2. \tag{5.28}$$

In Eq. (5.28), c_1 is the constant component for constant losses and auxiliary power requirements, c_2 is the coefficient for current-dependent losses, while i_{s} is the current at the FESS terminals, and c_3 is the coefficient for speed-dependent losses. The parameters c_1, c_2, and c_3 are estimated using real measurements from the FESS. The experimental setup and measurements are described later in Chapter 6. A simple curve fitting approach based on least-square error method is used to estimate the parameters of Eq. (5.28), and their values are given in Table 5.6. These values are estimated based on the difference between the power that the flywheel delivers and the power at the FESS terminals. It can be seen from Table 5.6 that the constant component of Eq. (5.28), which mainly consists of the auxiliary power required to run the system, is significant, considering a 60 kW FESS. However, it should be mentioned that some of the auxiliary components are meant for multiple flywheels, and the required auxiliary power for running multiple flywheels does not significantly increase with the number of flywheels.

Table 5.6: Parameters of the total system losses and auxiliary power requirements.

Parameter	Symbol	Value
Constant component	c_1	4.06 kW
Current-dependent component	c_2	$0.112\ \mathrm{kW/A^2}$
Speed-dependent component	c_3	$1.306 \times 10^{-4}\ \mathrm{kWs^2}$

5.3 Real-time Simulation of the High-speed FESS

In this section, the model of the high-speed FESS, described in section 5.2 using the model of each of its component, is simulated in real-time, and the results are compared with offline simulation results. For the real-time simulations, the model has been implemented in Opal-RT's RT-LAB real-time simulation environment. The real-time simulations are carried out on the Opal-RT's 5600 digital real-time simulator. RT-LAB generates a precompiled C code from the model on a host PC and loads this code on the real-time target. For offline non-real-time simulations, the ordinary differential equations of each component of the FESS, given in section 5.2, are discretized and solved using numerical methods in a script written in MATLAB programming language. The same simulation time step of 10 μs has been used in both cases. The simplified power system shown in Figure 5.12 is used to demonstrate the performance of the FESS. The equivalent grid and the transformer impedance are based on the values from the CIGRE European low voltage benchmark grid, in which a short-circuit capacity of 100 MVA and an X/R ratio of 1 has been considered for the medium voltage grid [123]. The load is a simple RL load. The model comparison is made using steps in the grid frequency and voltage.

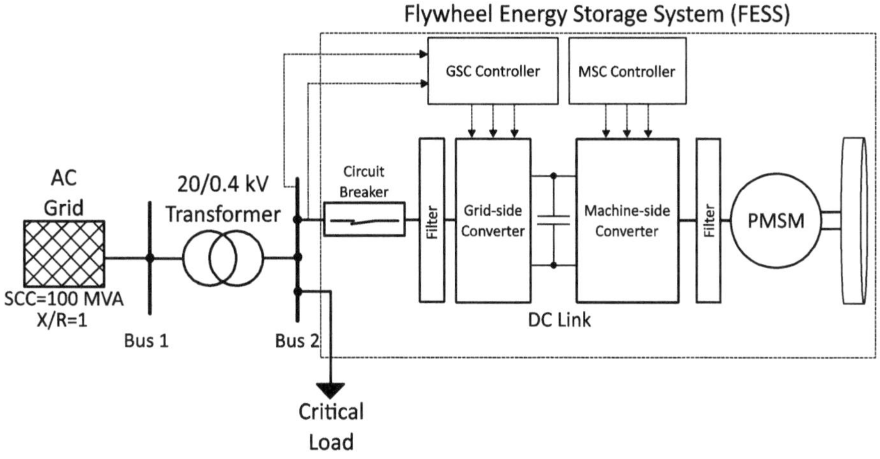

Figure 5.12: The single-line diagram of the simulated system.

5.3.1 Frequency Step Response

To show the fast dynamic behavior and full response of the FESS, a step of 1.2 Hz in the grid frequency is simulated, from 50 Hz to 48.8 Hz to fully observe the flywheel dynamics. Since the FESS is connected to a stiff grid, it cannot influence the frequency. However, it should provide active power to the grid according to its droop characteristic. In this case, the FESS should inject 100 % of its nominal power to the grid, according to the latest German grid code [182]. The high-speed FESS is initially charged to reach 36,000 rpm, with no power being exchanged with the grid prior to the simulated event.

The active power of the FESS model in real-time and offline simulation is shown in Figure 5.13, while other variables of the FESS are shown in Figure 5.14. The step in the frequency reference is applied at 0.04 s after the start of the simulation. As seen, the FESS quickly reaches its nominal power of 60 kW, and its rotational speed, shown in Figure 5.14(b), and therefore, its energy content begins to fall. Since only a short time span of 0.16 s is shown here to show the fast system dynamics, the drop in the flywheel speed is not significant. The DC-link voltage, shown in Figure 5.14(c), is quickly restored to 720 V by transferring the power injected from the PMSM to the grid by the grid-side converter. The electrical torque of the machine and the q-axis current required for providing this torque are depicted in Figure 5.14(d) and (f), respectively. As seen, a good match between the real-time simulation results obtained from the RT-LAB simulation software and the offline non-real-time results of the MATLAB script can be observed in all FESS variables. Two other steps in the grid frequency with amplitudes of 49.3 Hz and 49.75 Hz have also been applied for testing the response of the FESS model. The active power of the FESS from the real-time and non-real-time simulations for these steps in the frequency are also shown in Figure 5.13. It can be seen that in all cases the real-time simulation results in RT-LAB match the offline results for the MATLAB script.

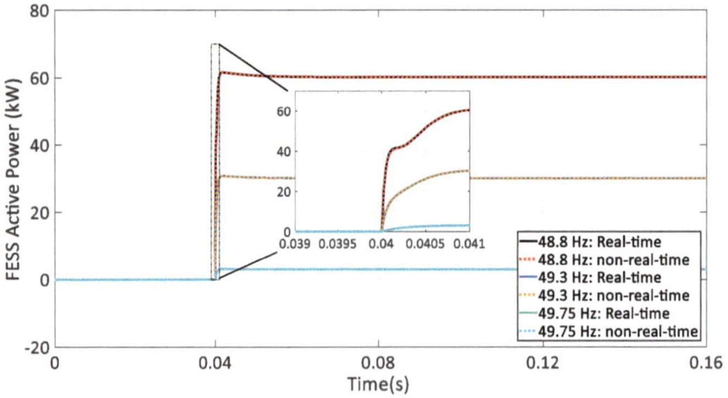

Figure 5.13: The active power injected by the FESS during different steps in the grid frequency

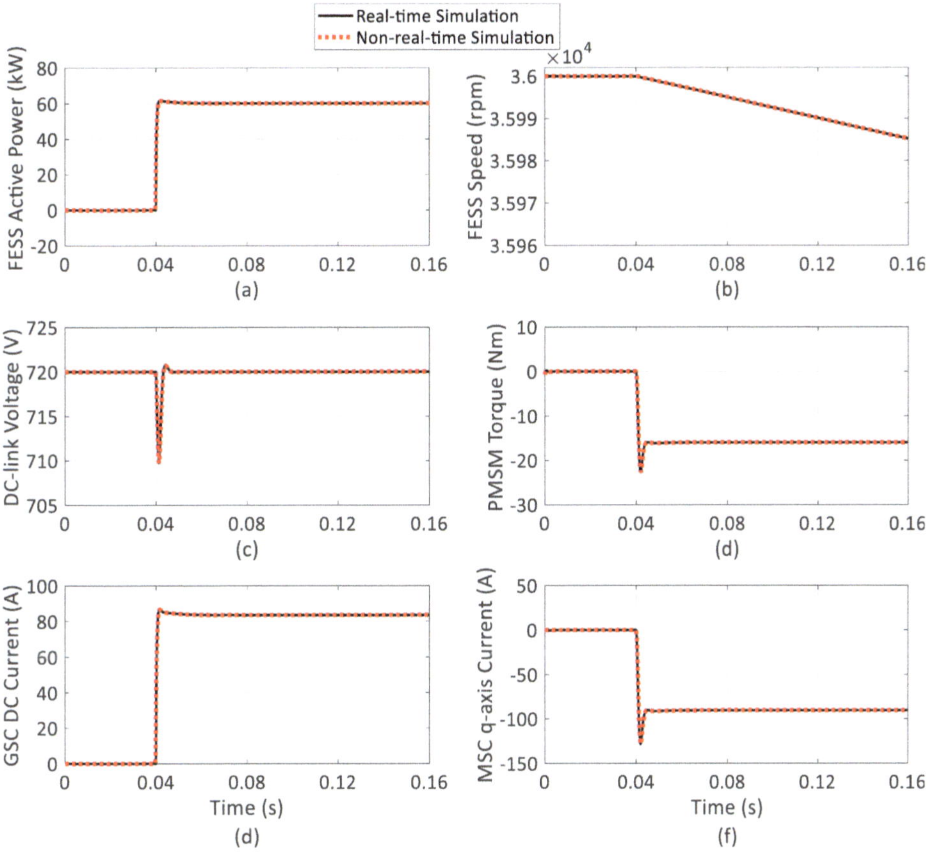

Figure 5.14: Real-time and non-real-time simulation results during a step in the grid frequency by 1.2 Hz.

5.3.2 Voltage Sag Response

In this scenario, a voltage sag on the medium voltage grid is simulated. Such voltage drops are common and could be caused by faults in the high voltage grids or parallel feeders [254]. The depth of the voltage sag is assumed be 7 % of the nominal values. During such a disturbance, the FESS should inject its maximum reactive power to the grid. The real-time and offline simulation results for this scenario are shown in Figure 5.15. The grid-side converter of the FESS injects approximately 50 kVar of reactive power, which the converter's maximum reactive power, to support the load during the voltage disturbance, as seen in Figure 5.15(a). This is done by injecting current in the q-axis direction, which is shown in Figure 5.15(e), as the reactive power is controlled by q-axis current of the grid-side converter. Since only reactive power is provided by the FESS, there is almost no change in the FESS speed, shown in Figure 5.15(b), the DC-link voltage, shown in Figure 5.15(c), and other variables of the machine and machine-side converter. Some minor transients are observed only at the beginning of the voltage sag. The speed of the FESS is slightly decreasing due to the losses within the FESS and the auxiliary power requirement. More importantly, the real-time simulation results match the non-real-time offline simulation in both steady-state and transient conditions for all variables of the FESS.

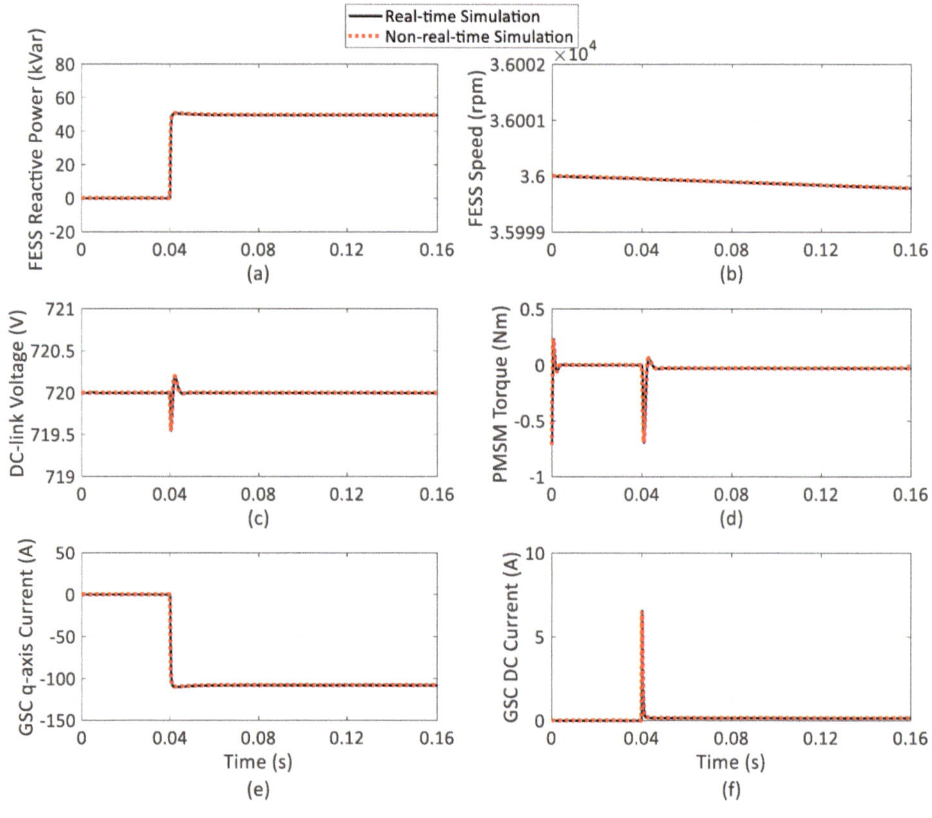

Figure 5.15: Real-time and non-real-time simulation results during a 7% voltage sag.

6 PHIL Testing and Model Validation of a High-Speed FESS for Frequency Support

This chapter presents the Power Hardware-in-the-Loop (PHIL) testing results of a 60 kW high-speed FESS for the application of frequency support. Firstly, the characteristics of the FESS, as the hardware under test, are given, followed by a description of the PHIL testing facility built and used for verifying the performance of the FESS. Next, several frequency deviation scenarios are simulated in real-time in order to evaluate the response of the FESS during such disturbances. The PHIL testing results are also used to validate the model of the high-speed FESS, described in Chapter 5. The same inputs which have been sent to the real FESS during the PHIL tests, are also given to the model of the FESS, running in real-time, and its response is compared with the response of the real FESS.

As discussed in Chapter 3, no PHIL testing of a FESS has been reported previously for verifying its response during frequency disturbances. There are only some limited experiments carried out for the application of supporting pulsed power loads in DC grids of shipboard power systems [24], [25]. Regarding the model validation of a FESS, an experimental validation of a FESS model has been reported in [255]. However, the tested system is a laboratory-scale 5.5 kW low-speed FESS with a maximum rotational speed of 3000 rpm, which is not a practical system with adequate capacity and power that can be used for ancillary services in the grid. In addition, no frequency control loop was implemented and tested. Therefore, this chapter presents the first PHIL testing of a high-speed FESS for frequency support, and the first model validation of a high-speed FESS.

6.1 Description of the Hardware under Test

The hardware under test is the 60 kW high-speed FESS, which has been modeled in Chapter 5. The inside view of the container of the FESS with all the main and auxiliary components is shown in Figure 5.2. This FESS has a maximum rotational speed of 45,000 rpm, which corresponds to a maximum capacity of 3.6 kWh, taking into account the flywheel's inertia. Therefore, this FESS can provide a discharge time of several minutes, depending on the discharge power. The high rotational speed of the FESS is achieved with the use of a carbon fiber reinforced plastic rotor, which is rotated using active magnetic bearings in a vacuum enclosure. The vacuum pressure of the flywheel containment is below 0.03 mbar during normal operation of the system. The flywheel steel containment is mechanically locked to a large underground cement foundation for safety purposes.

6.2 Description of the PHIL Setup

The PHIL testing of the FESS is conducted at the KIT's 1 MVA PHIL testing facility, as part of the EnergyLab 2.0, which is described in section 3.3. A schematic diagram of the setup tailored for the PHIL testing of the FESS is shown in Figure 6.1.

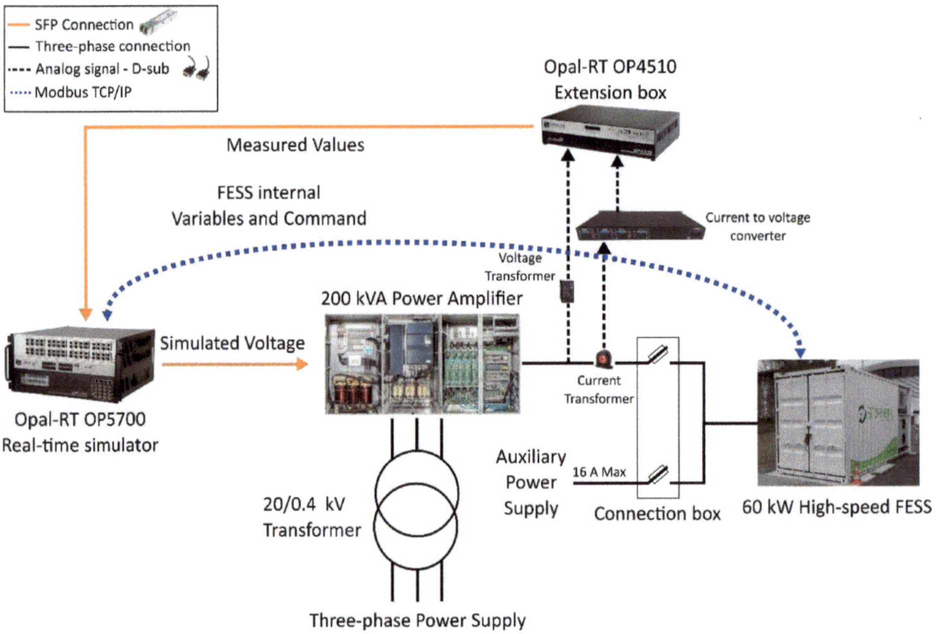

Figure 6.1: The schematic diagram of the setup used for the PHIL testing of the 60 kW high-speed FESS.

An Opal-RT 5700 real-time simulator is used for simulating the grid models and scenarios in real-time, controlling the power amplifier, collecting and recording measurement data, and monitoring and controlling the FESS operation during the experiments. The grid simulations, data collection, and the data conversion required for the Small Form-factor Pluggable (SFP) connection to the amplifier (see subsection 3.3) is carried out on one CPU core. A separate core of the simulator is dedicated for an Ethernet connection between the Programmable Logic Controller (PLC) of the FESS and the real-time simulator using the Modbus TCP/IP protocol. This is used to read the internal variables from the FESS controller and to send control commands to it. A list of the most important commands and measurement values communicated through the Ethernet connection with the FESS controller is given in Table 6.1.

Table 6.1: List of the most important commands and measurement values communicated through the Ethernet connection using the Modbus TCP/IP protocol between the FESS controller and the real-time simulator.

Signal Name	Unit	Data Type	Signal Name	Unit	Data Type
Commands			Measurements		
Operation mode command	-	U16	Active power measurement	W	U16
Active power reference	W	S32	Reactive power measurement	Var	S32
Reactive power reference	Var	S32	System state of charge	%	S32
Frequency reference	Hz	U32	Grid frequency	Hz	U32
Voltage reference	V	U32	Grid voltage	Hz	U32
Primary frequency control droop coefficient	W	U32	Kinetic Energy	Wh	U32

A single 200 kVA GAMP6 switched-mode power amplifier from Egston Power company is used for energizing the FESS using a four-wire three-phase cable. The power amplifier receives the voltage set points from the real-time simulator using a dedicated SFP connection. The voltages and currents of the FESS are measured at the FESS terminals using voltage and current transducers, and are sent to two separate Opal-RT OP4520 Kintex7 FPGA & I/O expansion units, due to the long physical distance between the hardware under test and the real-time simulator. Current and voltage measurements have been recalibrated before the experiments using a precise power supply.

Before and after the experiments, the vacuum, cooling water pumps, and several other auxiliary components of the FESS have to run continuously. Therefore, a manual switch was installed at the FESS terminals, which can switch the supply of the FESS from the power amplifiers to a conventional power supply, when no experiments are conducted.

The voltage-type ideal transformer method [256], which has been described in detail in subsection 3.2.2.1, is used for interfacing the hardware under test and the real-time simulator. For the current feedbacks a first-order low-pass filter with a cut-off frequency of 100 kHz is applied. A stable PHIL setup was observed using the ideal transformer method during all experiments and scenarios. Therefore, there has been no need to apply more complex PHIL interfacing algorithms or reduce the cut-off frequency of the filter.

6.3 PHIL Testing Scenarios and Model Validation of the FESS

The high-speed FESS is tested in the following scenarios:

- **Scenario A:** Charging the FESS at its rated power from standstill.
- **Scenario B:** Grid code compliance verification test for frequency support, according to the latest German grid code for low voltage grids.
- **Scenario C:** Response to the UK's frequency incident of 9 August, 2019.
- **Scenario D:** Increasing the frequency support contribution of a real low voltage grid in southern Germany.

The voltages, commands, and other input signals sent to the real FESS during the PHIL tests in each scenario are also given to the FESS model, running in real-time, for model validation. In the following, the PHIL testing results together with the response of the model of the FESS are provided for each scenario, and the response of the FESS and its model are compared with each other.

It should be noted that in all cases the FESS is tested in a closed loop with the current feedbacks being injected into the real-time simulation.

6.3.1 Scenario A: Charging at the Rated Power

This scenario includes charging the FESS from being completely discharged at a standstill condition to reaching its maximum rotational speed and the nominal capacity. The aim of this test is to derive the charging profile of the FESS, in particular at low rotational speeds, where the torque limitations of the machine prohibit the FESS from absorbing the nominal active power.

A simplified power system consisting of a voltage source and equivalent impedance with the values from the CIGRE European low voltage grid benchmark [123] is simulated in real-time on the Opal-RT's OP5700 real-time simulator. The voltages after the equivalent impedance are sent to the power amplifier, which feeds the FESS. Next, a step in the active power reference of the FESS from 0 to -60 kW is applied from the OP5700 real-time simulator using the Ethernet connection with the FESS controller and the Modbus protocol. The FESS should begin charging itself. When the FESS is fully charged and reached a state of charge of 100 %, which corresponds to a rotational speed of 45,000 rpm for this particular FESS, it should automatically stop absorbing power.

The results obtained from the PHIL testing of the FESS and the real-time simulation of its model in this scenario are depicted in Figures 6.2 and 6.3. The active power reference of -60 kW is applied at 5.7 s after the start of the test. As expected, the FESS can only absorb the nominal power above a certain rotational speed, due to the machine's torque limitations. This can clearly be seen in Figure 6.2, in which the ramp rate of the active power is limited, until reaching a state of charge of 24.6 %, which corresponds to a rotational speed of approximately 23,200 rpm. After exceeding this point, the FESS can absorb the rated active power until being fully charged. At the end, when the FESS is fully charged, the active power of the FESS drops to the minimum power required to cover the auxiliary power and losses in the whole system, as seen in Figure 6.2. This minimum power is required to maintain the state of charge at the maximum value.

It can be seen from Figures 6.2 and 6.3 that the real-time simulation model can reflect the charging profile of the real FESS with an acceptable accuracy. The time required to fully charge the FESS is approximately the same in both simulation and experimental results and the variations in the state of charge of the FESS during the charging are very close, with a maximum difference of just below 0.8 %. This small error could be due to the variations in the auxiliary power of the FESS during the experiment.

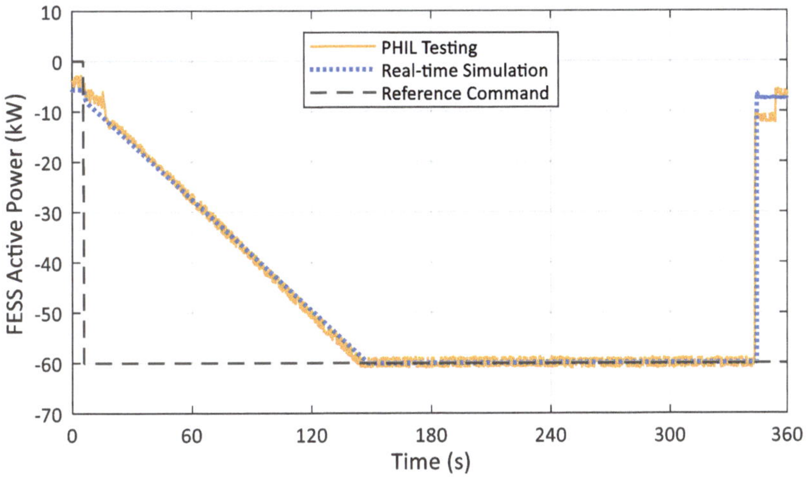

Figure 6.2: The measured and simulated active power of the FESS during the PHIL testing of scenario A, charging the FESS at the rated power.

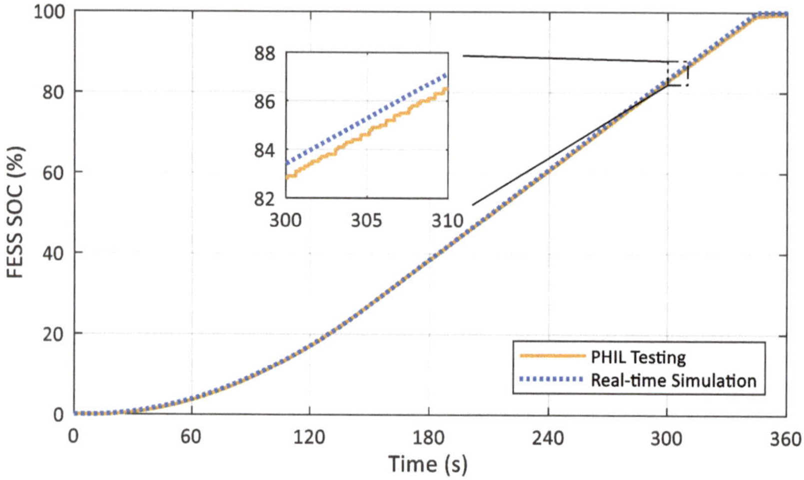

Figure 6.3: The measured and simulated state of charge of the FESS during the PHIL testing of scenario A, charging the FESS at the rated power.

6.3.2 Scenario B: Grid Code Compliance Verification Test for Frequency Support

The aim of this test is to verify that the high-speed FESS follows the P(f) characteristics during frequency deviations, according to the latest German grid code for connections to the low voltage distribution grids, the VDE-AR-N 4105:2018-11 [182]. The P(f) characteristic, shown in Figure 4.5(a), indicates the frequency deadband and different droop coefficients for over- and under-frequency incidents. According to this grid code, for frequency deviations above 200 mHz, the FESS should react within 2 s and reach the active power set point given by the P(f) characteristic with a rise time of under 2 s and a settling time of under 20 s. The acceptance tolerance for the difference between the active power set point and the measured active power is ±10 % [257].

In order to verify the grid code compliance of the FESS, a simplified power system similar to the previous scenario is simulated in real-time on the OP5700 real-time simulator. After charging the FESS, the frequency of the simulated grid is altered according to Figure 6.4. This test is adopted from the VDE V 0124-100:2020-06 [257], which describes the testing protocols for the generation units and storage systems that are planned to be connected to low voltage distribution grids. Conducting these tests is now a mandatory requirement for any ESS or generation unit, prior to a grid connection [257]. In comparison to the official testing requirements of the grid code, the tests have been shortened, taking into account the limited energy content of the FESS. It can be seen from Figure 6.4, that this test includes steps of different amplitudes in the emulated grid frequency. Each step in the frequency is kept for the period of 30 s, after which the frequency is set back to the nominal value. Each over-frequency step is followed by a subsequent under-frequency step with the same amplitude.

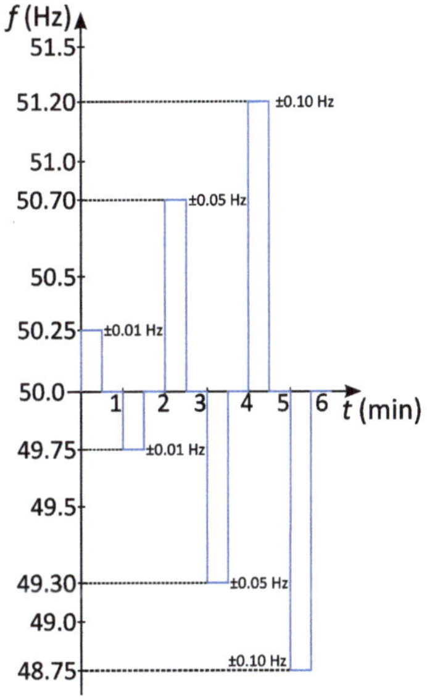

Figure 6.4: The variations in the grid frequency for grid code compliance validation of the high-speed FESS, adopted from the grid code VDE V 0124-100:2020-06 [257].

The tests are conducted with the FESS operating at the power factor of unity, as requested by the testing protocol [257]. The test is carried out after charging the FESS to reach a state of charge of 75 %, which corresponds to approximately 39,000 rpm for this FESS. At this state of charge, there are no speed-dependent ramp rate limitations due to the machine's maximum torque, and the FESS will provide the highest possible ramp rate, taking into account other constraints of the system, such as the measurement and communication delays.

The PHIL testing results of the 60 kW FESS and the real-time simulation results for this scenario are shown in Figures 6.5-6.7. The measurements from the FESS are obtained from the FESS internal controller, which shows some discontinuity in the results, due to the use of the Modbus protocol. It is clear from Figure 6.5 that the FESS controller can easily track the changes in the grid frequency with a high accuracy during frequency deviations. More importantly, it is shown in Figure 6.6, that the FESS quickly reacts to frequency changes larger than the frequency deadband, while ignoring smaller changes in the frequency. In the case of over-frequency steps, the FESS absorbs power from the grid with the rate of 40 % per Hz, after deducting the frequency deadband. For instance, it can be seen that the FESS absorbs 24 kW of power, when the emulated grid frequency is set to 51.2 Hz. During an under-frequency event, the FESS injects power to the grid with the rate of 100 % per Hz, after deducting the frequency deadband. As an example, it can be seen that when the simulated grid frequency is set to 49.3 Hz, the FESS injects 30 kW to the grid, which is 50 % of its nominal power. Therefore, it can be concluded the FESS accurately follows the P(f) characteristic required by the German grid code for an ESS connected at the low voltage level.

The simulation model of the FESS also follows the P(f) characteristics of the grid code, and a good match between the simulated power and the measured power of the FESS is observed in Figure 6.6. Therefore, the model of the FESS can accurately generate the same output of the real hardware. This is also shown in Figure 6.7, where the state of charge of the FESS is depicted. Between the state of charge of the real system and the simulation model, a maximum mismatch of only 0.35 % is observed in this scenario.

Furthermore, the fact that the FESS remains connected throughout these extreme frequency deviations and is not automatically disconnected from the grid by its protection systems demonstrates the frequency ride-through capability of the FESS according to the requirements of VDE-AR-N 4105:2018-11 [182].

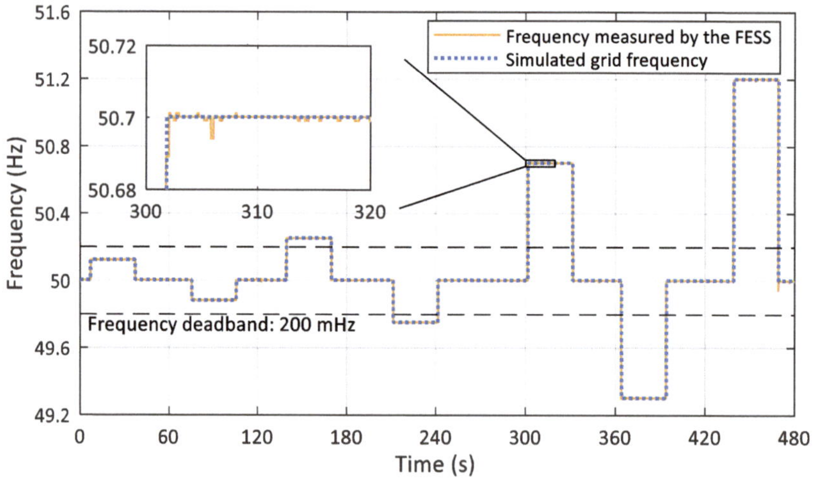

Figure 6.5: The measured and simulated grid frequency during the PHIL testing of scenario B, grid code compliance verification test.

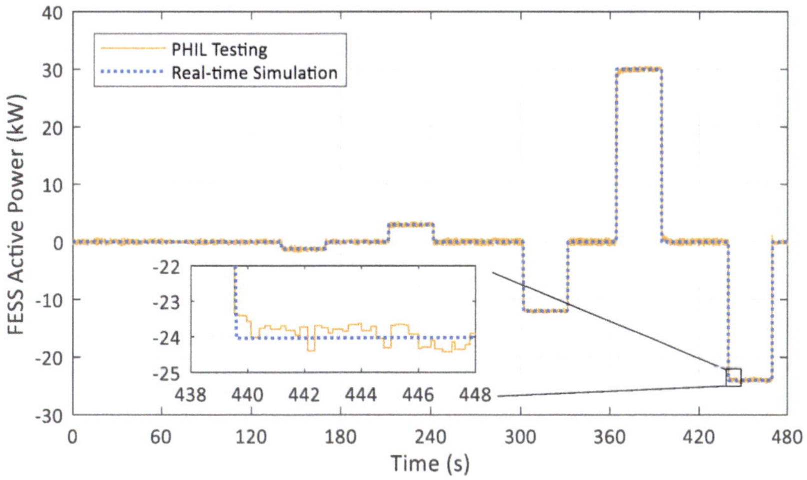

Figure 6.6: The measured and simulated active power of the FESS during the PHIL testing of scenario B, grid code compliance verification test.

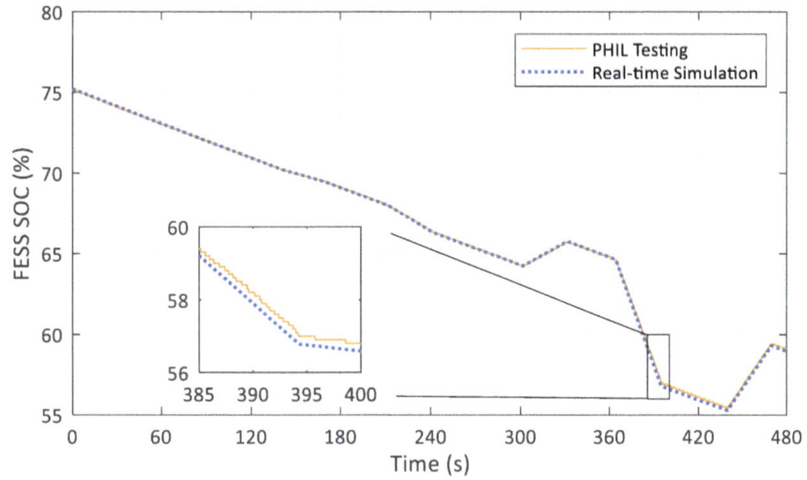

Figure 6.7: The measured and simulated state of charge of the FESS during the PHIL testing of scenario B, grid code compliance verification test.

In order to see how fast the real FESS responds to the grid frequency changes, one of the emulated steps in the grid frequency is shown more closely in Figure 6.8. Here, the active power is calculated directly from the voltage and current measurements at the FESS terminals, since the measurements from the FESS inertial controller have a high latency. It can be seen from Figure 6.8 that the FESS reaches the required active power in

58.5 ms, and settles in 88.6 ms. This is a very fast response time from the FESS. It is important to note that this time includes the time required for the phase-locked loop of the FESS to measure the grid frequency, the delay in signal transmission, and more importantly, the delay caused by the filter applied on the active power measurement. Therefore, the actual response time of the FESS can be even faster. This is by far faster than the requirements of the grid code in Germany, which requires an ESS to react to the frequency incident in below 2 s, to reach the set point in another 1 s, and to settle in below 20 s [257]. This delay is an important parameter, which is commonly neglected in studies regarding fast frequency support of converter-interfaced systems in power systems [12].

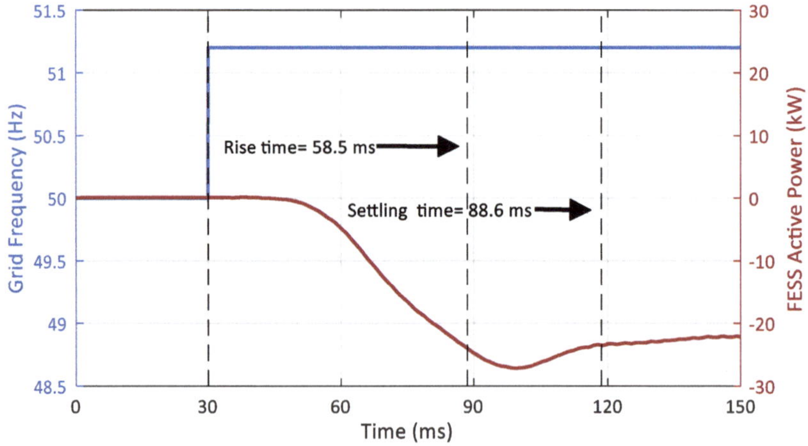

Figure 6.8: The response time of the high-speed FESS to a step in the grid frequency during the PHIL testing.

6.3.3 Scenario C: Response to the UK's Frequency incident of August 9, 2019

In this scenario, the recorded frequency variations during the major grid disturbance of August 9, 2019, in the power grid of the UK is simulated in real-time in order to evaluate the response of the high-speed FESS during such frequency deviations.

On this day, the UK's power system experienced a severe frequency drop, which led to the loss of power for approximately 1.1 million customers, including hospitals, airports, and railways [258]. The sequence of events, which is shown in Figure 6.9, was triggered by a lightning strike on a transmission line. The fault caused by the lightning strike resulted in a generation loss of approximately 1131 MW, which included 150 MW of distributed energy resources connected to the low voltage distribution grids, and reduction in the power generated by a wind farm by 737 MW. This significant loss in generation led to a severe drop in the grid frequency, reaching 49.1 Hz. As the frequency began to recover, another 210 MW gas turbine tripped, followed by an additional 350 MW power loss due to the activation of the Rate of Change of Frequency (ROCOF) protection system of several synchronous generators. The frequency dropped again, reaching a frequency nadir of 48.8 Hz. At this point, an under-frequency load shedding was enabled, disconnecting approximately 5 % of the total load, which included roughly 1000 MW of the loads at the low voltage distribution grids. With the activation of the under-frequency load shedding scheme, the frequency was regulated back to 50 Hz. The changes in the grid frequency in the power system of the UK during the time span of these incidents, which lasted approximately around 5 minutes, are shown with the blue dotted line in Figure 6.10.

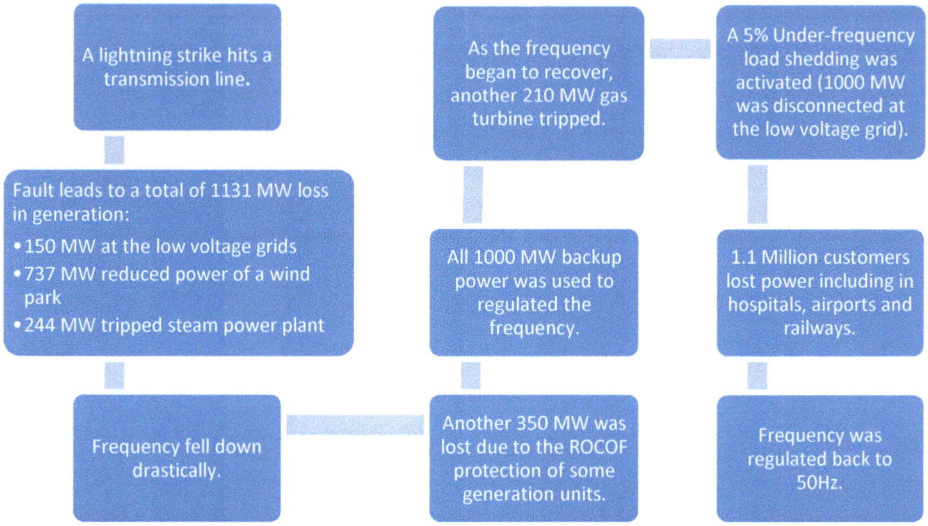

Figure 6.9: Sequence of event involved in UK's major frequency incident of 9 August, 2019 [258].

The original measured frequency data from the National Grid of UK has been used for testing the response of the FESS [259]. A simplified power system similar to the previous systems is simulated in real-time, where the frequency can follow the recorded frequency of the incident, when required. The FESS droop and deadband settings of the FESS are kept according to the P(f) characteristics of the German grid code VDE-AR-N 4105:2018-11 [182]. Prior to activating the frequency disturbance, the FESS is charged to reach a state of charge of approximately 95 %, which corresponds to the rotational speed of approximately 43,860 rpm.

The PHIL testing results of the 60 kW high-speed FESS and the real-time simulation results of its model during this scenario are shown in Figures 6.10-6.12. It can be observed from Figure 6.10, that similar to the previous scenario, the controller of the high-speed FESS easily tracks the changes in the grid frequency. Also, the FESS injects power to the grid, as soon as the frequency violates the frequency deadband of 200 mHz, and according to its droop setting, as seen in Figure 6.11. Since the maximum frequency deviation is approximately 1 Hz after the frequency deadband, the FESS will reach its maximum power of 60 kW at the minimum point of the frequency.

In addition, an excellent match between the active power of the real FESS and its model is observed in Figure 6.11, which indicates that both the real hardware and its model can successfully track the active power reference generated from the P(f) characteristics of the grid code. More importantly, the state of charge of the real hardware and its model during the time period of the disturbance are compared with each other in Figure 6.12. It can be seen, that there is a good match between the state of charge of the real FESS and the model in this scenario as well. There is a slight difference between the state of charge values of maximum 0.7 %, which could have been caused by the variability of the auxiliary power of the real FESS, such the power required for running the pumps and the cooling systems. It is also interesting to observe that the high-speed FESS has enough capacity to provide the required support throughout the whole frequency incident.

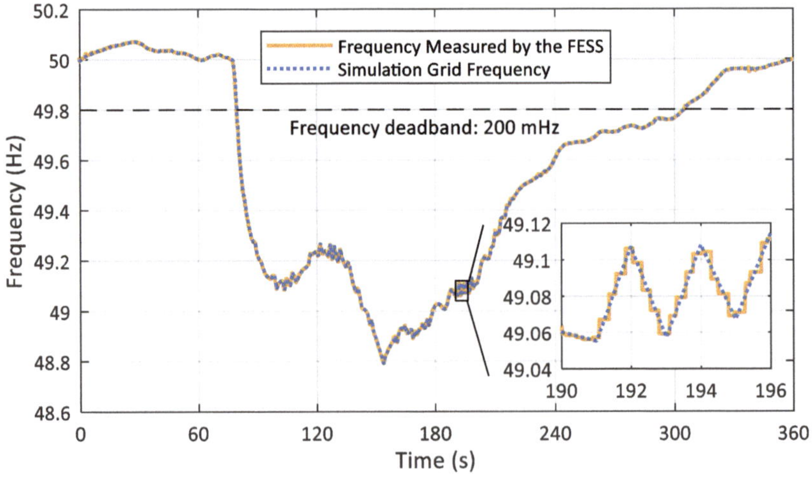

Figure 6.10: The measured and simulated grid frequency during the PHIL testing of scenario C, response to the frequency incident of August 9, 2019 in the UK.

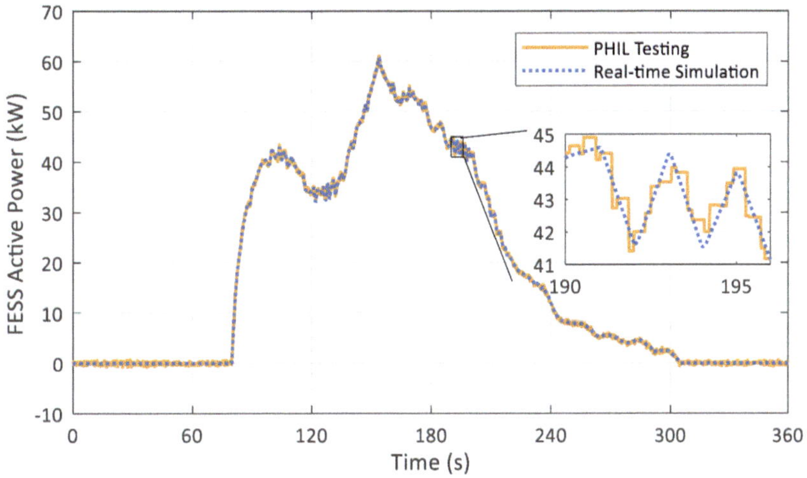

Figure 6.11: The measured and simulated active power of the FESS during the PHIL testing of scenario C, response to the frequency incident of August 9, 2019 in the UK.

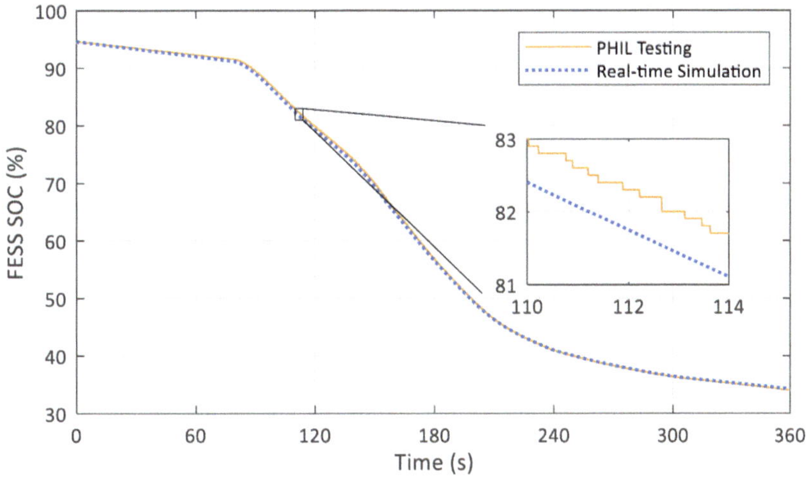

Figure 6.12: The measured and simulated state of charge of the FESS during the PHIL testing of scenario C, response to the frequency incident of August 9, 2019 in the UK.

6.3.4 Scenario D: Increasing the Frequency Support of a Low-voltage Distribution Grid

The goal of this scenario is to show the increase in the contribution of a low voltage distribution grid for supporting the grid frequency by using a FESS during frequency incidents.

A real low voltage distribution network in southern Germany is modeled and simulated in real-time in a collaboration with the distribution system operator. The single-line diagram of the modeled grid is depicted in Figure 6.13. This low voltage grid includes 78 short underground cables, 35 load connection points, and 3 PV systems. One of the PV systems is a medium-scale 100 kW PV system, while others are small residential ones with ratings below 10 kW. The medium voltage grid has an equivalent short-circuit capacity of 42.2 MVA with an X/R ratio of 1.61. These values are calculated based on the model of the medium voltage grid provided by the system operator, using the network reduction tool and extended Ward method [260] in DIgSILENT PowerFactory. The transformer and line data are also provided by the system operator.

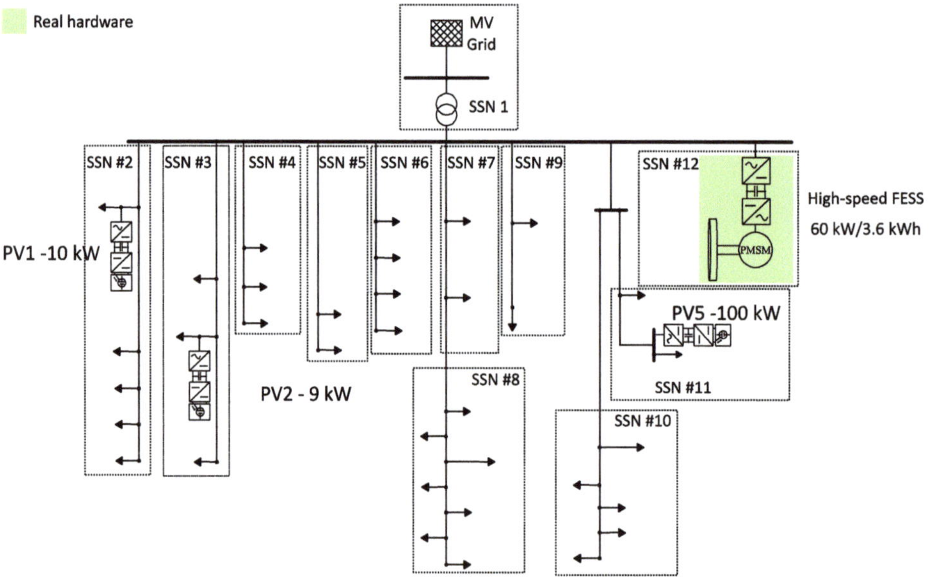

Figure 6.13: The single-line diagram of the real low voltage distribution network in southern Germany, used for PHIL testing of the FESS in scenario D.

The PV systems are modeled based on the generic two-stage PV model described in [261]. Conventional control loops including frequency control, active and reactive power control, DC-link voltage control, and maximum power point tracking using the incremental conductance method are also included in the PV models. For the PV cells, the parameters from of a commercial PV cell type, the Solarwatt Vision 60M style [262] are used.

Using conventional power system solvers such as state-space solvers, this network cannot be simulated in real-time with a simulation time step of below 200 μs, even without considering the PV systems. Therefore, it is not feasible to simulate accurately this grid model in real-time. This is due to the large number of short three-phase underground cables. Since the longest cable in this model is only about 300 m long, the conventional model decoupling method at long lines cannot be applied here, and one or several of the techniques listed in Table 3.1 have to be applied to be able to run this model in real-time.

Table 6.2 shows the required execution time for this grid model using different techniques listed in Table 3.1 to reduce the model execution time, which were calculated using the RT-LAB software. Firstly, the State-space Nodal (SSN) solver [109] is used to decouple the system equations into 12 subgroups, referred to as the nodal groups. This technique adds no artificial delays or impedance changes to the model, so it does not compromise the model accuracy. As shown in Figure 6.13, the secondary side of the transformer is selected as the main decoupling point (SSN node), as it generates a high number of groups from a single point. Using the state-space nodal solver, this grid model without the PV systems can be simulated in real-time with a simulation time step of 40 μs. Another method to reduce the model execution time is to neglect the line capacitances. This can be done for this particular grid as lines are mostly below 100 m long, are operated at the low voltage level, and the grid is assumed to be symmetrical [117]. This reduces the model execution time of the grid model without the PV systems to 20 μs, as it reduces the number of state variables by 8 for each line, and the grid under study has a relatively large number of lines. When compared to the model with the line capacitances, the grid model without the line capacitances shows on average a 0.06 % error in the steady-state currents and 0.14 % error in the steady-state voltages. By combining the use of state-space nodal solver and neglecting the line capacitances, the required model execution time for the grid without the PV systems decreases to only 8.72 μs. To add the PV models, the average converter models are used, as the study focuses on the frequency control [110]. By adding the PV models, the model execution time increases to 17.6 μs. Considering the added components to the model required for the communication to the power amplifier and with the two I/O expansion units, and also software safety features to shut down the system when extreme values are observed during PHIL testing, the simulation time step is set to 24 μs, in order to avoid any unexpected overruns. This relatively low simulation time step leads to a small loop delay. This improves the stability of the PHIL setup, as discussed in subsection 3.2.2.

The real-time simulation results of the grid model using state-space nodal solver are validated against offline results in MATLAB/Simulink, and no noticeable difference is observed. Moreover, the steady-state results are validated with the load flow results provided by the distribution system operator in PowerFactory, and the average difference between the two models was only 0.23 % in the steady-state voltages, and 0.3 % in the steady-state currents.

Table 6.2: Comparison of the model execution times of the German low voltage distribution grid using different techniques used for reducing the model execution time.

Model	Neglecting capacitances	Using SSN	PV included	Model execution time (μs)
Model 1	No	No	No	> 200
Model 2	No	Yes	No	40
Model 3	Yes	No	No	20
Model 4	Yes	Yes	No	8.72
Model 5	Yes	Yes	Yes	17.6

The real high-speed FESS has been integrated into the grid model at the secondary side of the main transformer, as depicted in Figure 6.13. A separate nodal group is dedicated for the connection of the FESS using the current values measured at the real FESS terminals.

A slow frequency ramp is simulated in the medium voltage grid with the frequency slowly increasing from 50 Hz to 51.2 Hz with a rate of 0.25 Hz/min. The real FESS together with the simulated PV systems react to the frequency ramp according to the German grid code for low voltage grids. The grid code indicates that both PV generation and storage units have to participate in the over-frequency incidents and reduce their output with the rate of 40 % per Hz, when the frequency passes the 200 mHz threshold.

The FESS is charged to approximately 71 % before conducting this experiment. The results of the PHIL testing of the FESS together with the results of the real-time simulation of the grid in this scenario are shown in Figures 6.14 and 6.15. At 10 s after the beginning of the test, the PV systems are switched on with an enabled maximum power point tracking at the irradiance density of 1 kW/m^2 and an ambient temperature of 25 °C. Therefore, they quickly reach their maximum power output, as seen in Figure 6.14(a). This reduces the total power that low voltage grid draws from the medium voltage grid from 140 kW to approximately only 25 kW. Such low powers are common around noon in low voltage grids with a high share of PV generation, as previously seen in Chapter 4. The frequency ramp is enabled at 27.4 s. As soon as the frequency passes the frequency deadband, the simulated PV systems starts to curtail their generation during the over-frequency event, as seen in Figure 6.14(a), and contribute to decreasing the total generation in the system. At the same time, the real FESS starts to draw active power from the grid according to its droop coefficient and begins to charge itself, as shown in Figure 6.14(d). This increases the consumption in the low voltage grid. When the frequency reaches 51.2 Hz, the total active power of the low voltage grid has increased from 25 kW to approximately 94 kW (seen Figure 6.14(c)). Approximately one third of the increase in power is the share of the real FESS, while the rest is due to the curtailment of the generation in the simulated PV systems. Therefore, the active components of the low voltage grid contributed to the regulation of the frequency of the whole system.

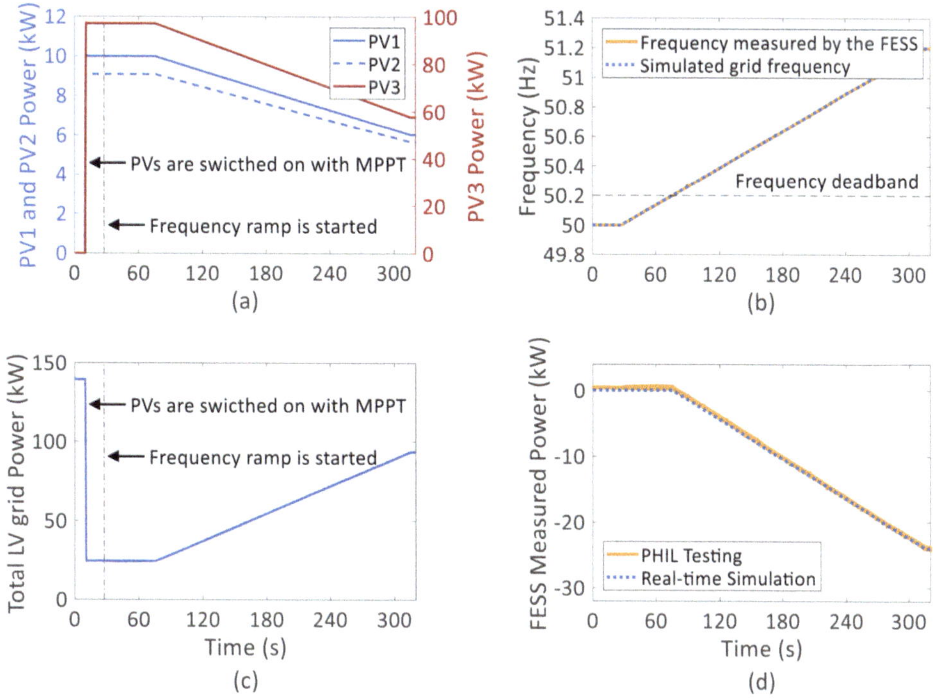

Figure 6.14: The PHIL testing results of scenario D, increasing the frequency support of low-voltage grids. (a) Power of the PV systems, (b) the measured and simulated grid frequency, (c) The total power drawn by the low voltage (LV) grid, and (d) the measured and simulated power of the FESS.

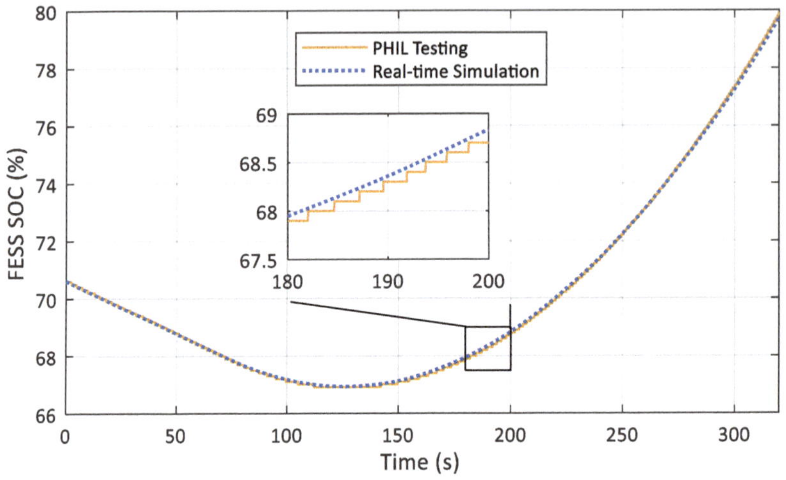

Figure 6.15: The measured and simulated state of charge of the FESS during the PHIL testing of scenario D, increasing the frequency support of low-voltage grids.

In addition, similar to the previous scenarios, a good match between the active power of the real FESS and the simulated one, running in real-time, is observed, as seen in see Figure 6.14(d). Also, it is clear from Figure 6.15, that the model can reflect the variations in the state of charge of the FESS with a high accuracy in this scenario as well.

6.4 Summary

This chapter presented the PHIL testing results of a 60 kW/3.6 kWh high-speed FESS in several frequency deviations scenarios. It was demonstrated the high-speed FESS can accurately follow the P(f) characteristics required by German grid code for low voltage grids during under- and over-frequency disturbances. Furthermore, it was shown that the FESS can reach the required power in under 60 ms during frequency disturbances and support the grid by quickly charging and discharging itself. Moreover, the nonlinear charging profile of the FESS from a standstill condition until reaching the maximum rotational speed was demonstrated, showing the limitations of a FESS at low rotational speeds.

In addition, it was shown that the model of the FESS, presented in Chapter 5, can approximately generate the same power and state of charge of the real FESS in real-time, during all the investigated scenarios.

7 Adaptive Inertia Emulation using High-speed Flywheel Energy Storage Systems

As discussed in Chapter 1, the continuous growth in the share of converter-interfaced renewable generation has led to a steady decrease in the total inertia in power systems. The reduction in the system inertia can lead to an increase in the Rate of Change of Frequency (ROCOF) and the severity of the frequency deviations during imbalances in generation and demand. Inertia emulation techniques using converter-interfaced systems can help to decrease the effects of low physical inertia in the system by providing synthetic inertia support during frequency disturbances.

In this chapter, a new adaptive inertia emulation controller for high-speed FESS is introduced. The proposed controller aims to reduce simultaneously the ROCOF and the frequency nadir by adaptively changing the inertia and damping coefficients of the controller. It uses a combination of the bang-bang control approaches and self-adaptive ones to benefit from the advantages of both control designs. The performance of the proposed adaptive controller is validated initially using offline numerical simulations in a low voltage microgrid. Next, the proposed controller is implemented on the 60 kW high-speed FESS, described in chapter 5 and 6, using the concept of rapid control prototyping. The performance of the FESS using the proposed inertia emulation controller is validated using PHIL testing of the FESS in the low voltage microgrid in two different scenarios. The simulation and PHIL testing results confirm that the use of the FESS with the proposed inertia emulation controller can effectively reduce the ROCOF and the frequency deviation, and the suggested adaptive control design outperforms several adaptive methods previously reported in literature.

7.1 Introduction

In this section, the need for inertia emulation using converter-interfaced systems is discussed briefly. The impact of the cumulative system inertia on the frequency dynamics of a power system during a power imbalance is demonstrated using a simplified single-machine power system. A comparison among the possible active power sources required for emulating the inertia response using converter-interfaced systems is also presented, aiming to justify the advantages of using a FESS for this application. Lastly, a literature review on different adaptive control methods for providing inertia support is provided.

7.1.1 The Need for Inertia Support from Converter-interfaced Systems

Inertia of a power system resists fast variations of its frequency during disturbances. Analyzing the previous major frequency disturbances in Europe reveals that a stable system operation can be guaranteed only up to a ROCOF of 1 Hz/s, which corresponds to a system imbalance ratio of 20 %. However, imbalance ratios of more than 40 % are expected for the near future in Europe, which leads to a ROCOF of more than 2 Hz/s [5]. In poorly interconnected grids, such as in Australia, a ROCOF as high as 6 Hz/s has already been experienced in 2016, which led to a widespread system blackout, due to the activation of under-frequency and ROCOF protection relays of the synchronous generators of conventional power plants [263].

Figure 7.1 illustrates a typical frequency behavior of a power system during a major loss in generation. The initial slope of the drop in the frequency, or the ROCOF, is limited only by the cumulative system inertia. The system inertia plays an important role in maintaining the grid stability [264]. The inertia also has an impact on the minimum frequency reached during the under-frequency event, which is called the frequency nadir [265]. A low-inertia power system can experience a high ROCOF and an extreme frequency nadir during disturbances.

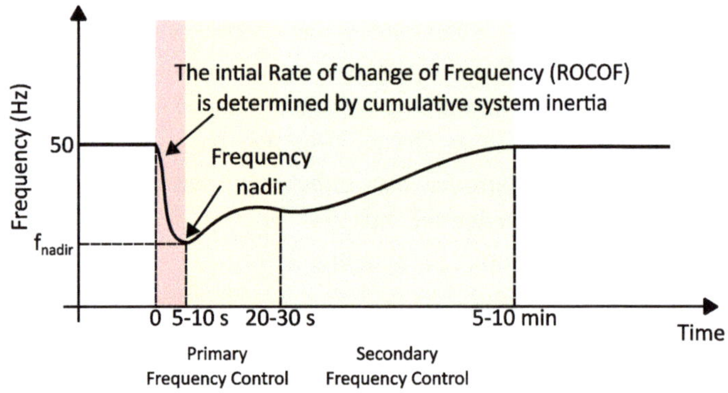

Figure 7.1: Typical behavior of the frequency in a power system after a major large loss in generation (adopted from [266]).

The impact of the system inertia on the frequency and the ROCOF can be seen more clearly if we simplify the frequency control loops of a power system to the block diagram shown in Figure 7.2, as proposed in [267]. In Figure 7.2, a multi-machine system is represented by an equivalent single machine, where the inertia constant H_g is the equivalent inertia constant of all generating units, and the damping coefficient D is the equivalent damping of all the generators and loads in the system. Moreover, it is assumed that all frequency-related control loops of the turbine-generators have the same structures and parameters. In this figure, ΔP_L is the changes in the loads, ΔP_m is the variations in the generator mechanical input, ΔP_c is the control power demanded by the secondary frequency control, Δf is the variation in the grid frequency, R is the droop coefficient, T_1, T_2, T_3 and T_4 are the turbine and governor time constants. The inertia emulation using an inverter-based system is added to this block diagram with the emulated inertia constant of H, and the converter represented by a simple time constant of T_c [110]. This time constant represents the delay and the bandwidth of the inverter. The parameters used for simulating this simplified model are given in Table 7.1.

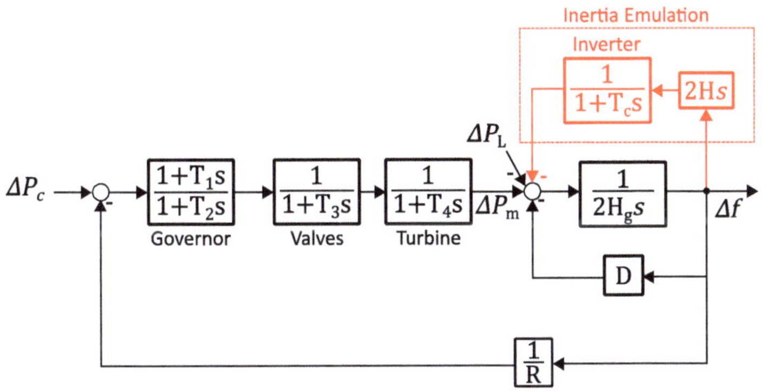

Figure 7.2: Simplified block diagram of frequency control loops in a simplified power system with the addition of emulated inertia (model according to [267]).

Table 7.1: The parameters used for the simplified single-machine power system.

Parameter	Symbol	Value
Generator inertia constant	H_g	3 s
Damping coefficient	D	1 p.u.
Governor droop	R	0.05 p.u. (5 %)
Inverter time constant	T_c	0.1 s
Governor lead time constant	T_1	0.2 s
Governor lag time constant	T_2	1 s
Valve time constant	T_3	0.5 s
Turbine time constant	T_4	5 s

The response of the simplified power system model of Figure 7.2 to a small step increase in the load (ΔP_L) is shown in Figure 7.3 for different values of the emulated inertia constant (H). As seen, the increase in the emulated inertia reduces the maximum ROCOF and limits the frequency nadir, at the price of a slower recovery of the grid frequency. The increase in the frequency settling time, after recovering from the frequency nadir, is considered the main drawback of increasing the system inertia using inertia emulation techniques.

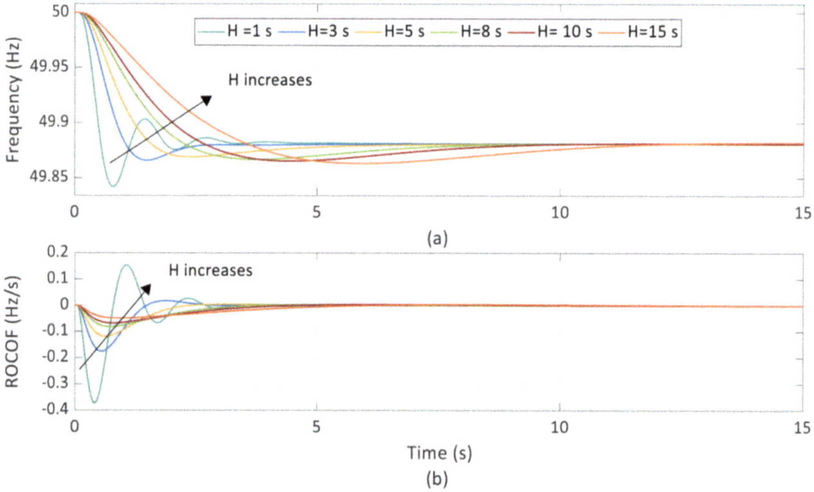

Figure 7.3: The effect of increasing the emulated inertia constant using converter-interfaced systems on (a) the frequency, and (b) the ROCOF during an under-frequency event.

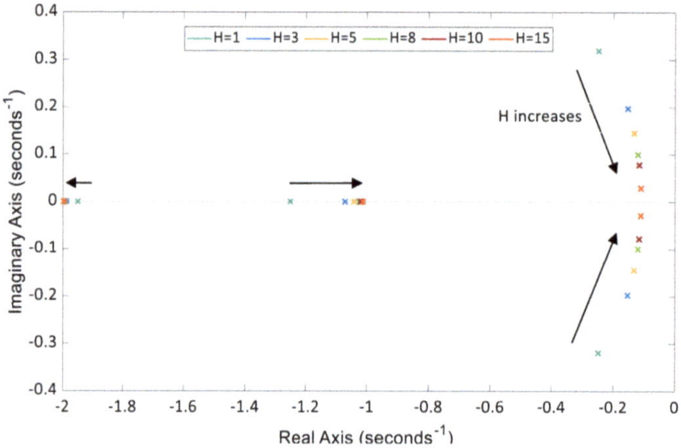

Figure 7.4: The pole-zero map of the simplified power system with the increasing value of the emulated inertia constant (H).

This can also be observed from the pole-zero map of the same system for different values of inertia, shown in Figure 7.4. With the increase in inertia, the dominant system poles move towards the origin, which indicate a slower decaying response and a higher settling time. The pole-zero map has been drawn by neglecting the inverter time constant (T_C), as it is much smaller than the time constants of the turbine and governor. However, unlike the physical inertia of synchronous generators, emulated inertia is a control parameter which can be altered in real-time according to the system requirements. Therefore, to solve the problem of slow frequency recovery, adaptive inertia controllers have been proposed [265], [268]–[271], where the emulated inertia is reduced to near zero after the frequency nadir is reached for a faster recovery of the frequency. These adaptive inertia emulation techniques are later discussed in detail in section 7.2.2.

It can be seen in Figure 7.3, that the inertia has no influence on the steady-state frequency value. The steady-state frequency depends on the system damping (D) and the droop (R). A similar analysis on the effect of increasing the damping, rather than the inertia constant, shows that increasing the damping improves the frequency nadir and reduces the steady-state frequency error. The frequency nadir is determined by both, the system inertia and the damping, with the damping having a greater impact [265]. It can be concluded, that increasing the inertia and the damping can reduce the ROCOF and frequency deviation during imbalances of generation and demand. However, the inertia constant should be reduced when the frequency is recovering.

7.1.2 Flywheels as the Active Power Source for Inertia Emulation

Initially proposed for wind turbines [272], the topic of providing inertia support using converter-interfaced systems has gained a great attention over the past decade. Significant contributions have been made by researchers working in this field, for instance in [273]–[281], where a variety of techniques and controllers has been proposed to emulate the inertia response of the synchronous generator using converter-interfaced systems during frequency deviations. Comprehensive reviews of many of these techniques can be found in [282] and [283].

Unlike the synchronous generators, where the kinetic energy stored in their large rotating rotors is used for providing the inertia response, converters do not inherently have a source of energy. Therefore, an active power source is required for providing the power demanded by the inertia emulation controller. In many of the proposed control designs for inertia emulation, for instance in [273]–[276], the active power source behind the inverter are simplified, assuming an ideal DC-link with an unlimited energy content.

However, in practice, the dynamics and limitations of the active power source can also influence the inertia response of such systems. When the active power source is included, wind turbines [278], [279], PV systems [280], [281], DC-link capacitors [229], [284], and batteries [285], [286] are suggested to be the source of energy for inertia emulation. However, as summarized in Table 7.2, each of these solutions have their own limitations in providing the power required for emulating the inertia response. These limitations can change the response of the system from an ideal inertia response. In wind turbines, the speed recovery required after the first seconds of providing the inertia response, alters the response of the wind turbine from an ideal inertia response [272], [287]. A poorly conducted speed recovery can even cause a second frequency drop. Furthermore, inertia emulation may cause rotor stall or even trigger instabilities if the aerodynamics of wind turbines are not fully studied. Photovoltaic systems cannot participate in under-frequency events when being operated at their maximum power point. Therefore, additional storage systems or operating the system below the maximum power point is required, which is not ideal from an economical aspect. In an attempt to solve this, authors have also suggested to use the DC-link capacitors of the inverters as the active power source for inertia support [229], [284]. However, considering the low capacity of these capacitors and the voltage constrains for maintaining a linear pulse width modulation, the amount of energy that it can be provided for inertia emulation using these capacitors is very limited. Battery energy storage systems can be a good alternative for inertia emulation [285], [286]. However, concerns over the lifetime of the battery may force the operator to limit its power ramp rate, which can impair the inertia response. Most adaptive inertia emulation techniques, like the methods proposed in [264], [265], [268]–[271], require a sudden burst of power for several seconds, which is not the ideal use of a battery, concerning its lifetime and capacity. Therefore, it has been suggested to combine the battery with a supercapacitor in a hybrid structure, where the supercapacitor is mainly responsible for the inertia support [288]. However, this design leads to a higher degree of complexity in both control and hardware requirements. Moreover, the emulated inertia is limited by the size of the supercapacitor.

As discussed in Chapter 2, a FESS can rapidly provide high amounts of power with a high ramp rate and no concern over its lifetime or capacity. The FESS does not have any of the aforementioned limitations and can provide an ideal inertia response. In a high-speed FESS, the high-inertia rotor can store a significant amount of kinetic energy. However, the high-inertia rotor is electrically decoupled from the grid by a back-to-back converter. Nonetheless, by implementing an inertia emulation controller on the converter, the inertia can be transferred artificially to the grid side, providing an inertia response, similar to the response of the synchronous generators. The inertia emulation controller for a FESS is described in detail in section 7.2.

Table 7.2: Technical limitations of different active power sources for inertia emulation.

Active Power Source	Literature	Technical Drawbacks/Limitations
Wind turbines	[272], [279]	The mandatory speed recovery greatly changes its response from an ideal inertia response [272], [287].
		Can cause rotor stall and system instability [272], [287].
PV systems	[280], [281]	Requires operating the system below maximum power point or additional storage systems for under-frequency events.
DC-link capacitors	[229], [284]	Very limited energy content for a single inverter and therefore, limited inertia.
		Constrained voltage deviation to guarantee linear pulse width modulation and avoiding overstressing of the semiconductors.
Battery energy storage system	[285], [286]	Concerns over battery lifetime can force to limit the power ramp rate.
Hybrid storage: Battery and supercapacitor	[288]	Complex design of the controllers and hardware requirements.
		The emulated inertia is limited by the size of the supercapacitor.
Flywheel Energy Storage System (FESS)	[45], [46]	None.

7.1.3 Inertia Emulation versus Virtual Synchronous Machines

There are two common control structures for providing inertia support using converter-interfaced systems. In one category, the virtual synchronous machines [289]–[291], a low- or high-order model of a synchronous generator is used to generate the reference values for the voltage amplitude and phase angle of a grid-connected converter. In this design, the active power is controlled by changing the voltage phase angle, while the reactive power is controlled by varying the voltage amplitude, similar to synchronous generators. The main advantage of this approach is that, in some designs, it can eliminate the need for a frequency measurement system such as a phase-locked loop [291]. It uses the phase angle from the generator equation for synchronization of the converter. However, implementing virtual synchronous machines requires a drastic change and redesign of the converter controllers. Therefore, it is not logical or practical to implement such controllers on existing converters with already built-in controllers [279]. Also, such controllers do not inherently have current controllers. Therefore, an additional protection system to limit the currents may be required [292]. Moreover, for providing the inertia support using virtual synchronous machines, the active power controller including the inertia controller should be implemented on the grid-side inverter. However, in most commercial FESS, similar to many wind turbines, the active power controller is implemented on the machine-side converter. This design choice allows to directly control the machine torque, and facilitates parallel connection of several flywheels to reach higher ratings.

In the second category, which is referred to as inertia emulation techniques, only the swing equation of the synchronous generator is used to generate the active power reference for the converter. Therefore, the inertia emulation controller can be easily added as an outer control layer to the conventional current-controlled voltage source converters, including existing systems [292]. Due to this major advantage, this control design has been used by the European VSYNC project [293], as well as in [273], [279], [294], [295]. This method is also applied in this work, as it aims to implement the inertia emulation control as an external control loop to a commercial high-speed FESS, with pre-existing converter controllers.

The drawback of this approach is the need for estimating the frequency and its derivative. In this work, the challenges associated with the use of a conventional phase-locked loop and the differential operator for estimating the frequency derivative are avoided by using a Dual Second-order Generalized Integrator Frequency-locked Loop (DSOGI-FLL). This is explained later in subsection 7.2.4.

7.1.4 State-of-the-Art in using a FESS for Inertia Emulation

The use of a FESS for inertia emulation has been previously reported in [45] only, while the work in [46] presents a virtual synchronous generator using a FESS. However, in both cases, the parameters of the proposed controllers for inertia support are constant and adaptive controllers are not applied. The controller in [46], as a virtual synchronous machine controller, has no current control and authors report problems in forming pure sinusoidal waveforms. In both studies, realistic grid scenarios are not provided, and the performance of the proposed controllers has been demonstrated using numerical simulations only. To our knowledge, no experimental validation of using a FESS for inertia emulation has been reported. Furthermore, using PHIL testing for the validation of inertia emulation techniques has been reported only in [273], but the tests include only testing the inverter with a simple control design, and an ideal DC voltage source has been used as the active power source. Therefore, PHIL testing of a full system, including the active power source for inertia emulation, has not been conducted before.

This work attempts to improve the current state-of-the-art on using a FESS for inertia emulation by introducing a new adaptive inertia emulation controller, implementing it on a real FESS, and validating it using PHIL testing. The proposed controller combines the advantages of bang-bang controllers with self-adaptive ones for an improved response to frequency disturbances. Moreover, it considers the state of charge of the FESS in providing the inertia response and considers the practical limitations of using a FESS for frequency support. The proposed controller is described in the following section.

7.2 Adaptive Inertia Emulation Control for a High-speed FESS

In this section, the proposed adaptive inertia emulation controller for high-speed FESS is described in detail. Firstly, the controller structure for inertia emulation using FESS is explained, followed by the description of the suggested adaptive control method to alter the inertia constant and damping coefficient for an improved frequency regulation. Next, the choice of the parameters for the proposed controller is explained, considering the practical limitations of a FESS. Lastly, the method used to estimate the frequency and its derivative, as the inputs of the proposed inertia emulation controller, is described briefly.

7.2.1 Controller Design of Inertia Emulation using High-speed FESS

From the mechanical model of a PMSM in a high-speed FESS, described in Chapter 5, the electrical power of the PMSM (P_e) is calculated as

$$P_e = \mathrm{J_f}\omega_\mathrm{m}\frac{d\omega_\mathrm{m}}{dt} + \mathrm{D_f}\omega_\mathrm{m}. \tag{7.1}$$

Equation (7.1) can be rewritten in per unit values using the definition of the inertia constant for a PMSM as [231]

$$\mathrm{H} = \frac{\frac{1}{2}\mathrm{J_f}\omega_\mathrm{max}^2}{\mathrm{S_n}}. \tag{7.2}$$

In Eq. (7.2), ω_max is the maximum speed of the FESS, and $\mathrm{S_n}$ is its nominal apparent power. The inertia constant H given in seconds determines the time the flywheel can provide the nominal power using the stored kinetic energy only. By indicating the per unit value with an upper bar, Eq. (7.1) becomes

$$\bar{P}_e = 2\mathrm{H}\frac{d\bar{\omega}_\mathrm{m}}{dt} + \mathrm{D}\bar{\omega}_\mathrm{m}, \tag{7.3}$$

where D is the damping coefficient in per unit. Assuming the number of pole pairs of the PMSM to be one, the mechanical and electrical angular velocity become identical, therefore,

$$\bar{P}_{\mathrm{e}} = 2\mathrm{H}\frac{d\bar{\omega}_{\mathrm{e}}}{dt} + \mathrm{D}\bar{\omega}_{\mathrm{e}}. \tag{7.4}$$

It is evident from Eq. (7.4) that the PMSM in a FESS is governed by the same swing equation of the conventional synchronous generators, with the difference of having no mechanical power input in a FESS. With any changes in the angular velocity of the supplied voltage to the PMSM, the PMSM provides an inertia response according to Eq. (7.4). However, in a FESS, the PMSM is decoupled from the grid by a back-to-back converter. To go around this problem, we replace the electrical angular velocity of the PMSM $\bar{\omega}_{\mathrm{e}}$ with the error in the grid's angular velocity in per unit $\Delta\bar{\omega}_{\mathrm{g}}$. The resulting equation is then used to generate the active power reference ($\bar{P}_{\mathrm{ref}}$) for the FESS. That is,

$$\bar{P}_{\mathrm{ref}} = 2\mathrm{H}\frac{d\Delta\bar{\omega}_{\mathrm{g}}}{dt} + \mathrm{D}\Delta\bar{\omega}_{\mathrm{g}}. \tag{7.5}$$

Equation (7.5) is the basis for the proposed inertia emulation controller. Using this approach, the inertia response of the PMSM is artificially transferred to the grid side, and the FESS reacts to the grid frequency deviations, similar to a synchronous generator.

The block diagram of the proposed inertia emulation controller is shown in Figure 7.5. As mentioned before, due to the structural design of commercial FESS, the active power control is often implemented on the MSC controller. The active power reference generated by Eq. (7.5) is first transformed to the electrical torque reference τ_{eref} using the instantaneous speed of the flywheel ω_{m}, which is then converted to the reference value for the q-axis current of the MSC using

$$i_{\mathrm{qref}} = \frac{2\tau_{\mathrm{eref}}}{3n_{\mathrm{p}}\psi_{\mathrm{PM}}} = \frac{2\bar{P}_{\mathrm{ref}}\mathrm{P_n}}{3n_{\mathrm{p}}\psi_{\mathrm{PM}}\omega_{\mathrm{m}}}. \tag{7.6}$$

The reference value for the d-axis current, the inner current controllers, and the GSC controller are identical to the ones presented in Chapter 5. The GSC controls the voltage of the DC-link according to the power injected or drawn from the MSC. Therefore, the power of the PMSM is transferred immediately to the grid. The GSC also regulates the voltage on the grid side, by controlling the reactive power according to the Q(U) characteristic of the German grid code, as shown in Figure 4.5(b).

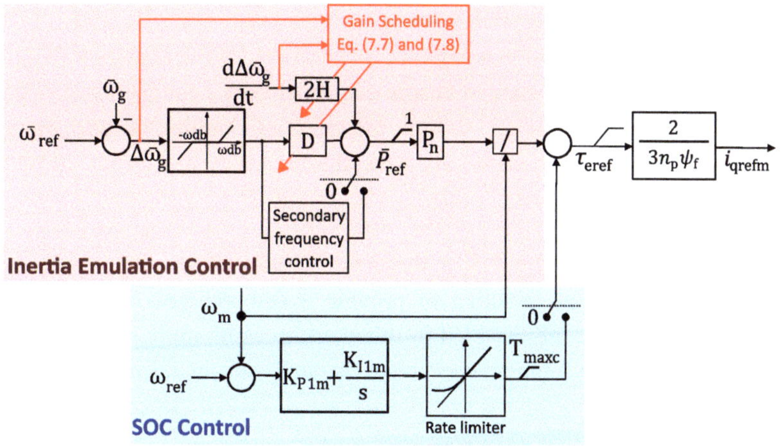

Figure 7.5: The block diagram of the proposed inertia emulation controller, and the State of Charge (SOC) controller. This controller is to be added the MSC controller of the FESS in Chapter 5.

It can be observed from Figure 7.5, that the inertia constant H determines the amount of active power injection with respect to the ROCOF ($\frac{d\Delta\bar{\omega}_g}{dt}$), while the damping coefficient D determines the active power reference according to the error in frequency itself ($\Delta\bar{\omega}_g$). As mentioned before, a major advantage of inertia emulation in comparison to the physical inertia is that the emulated inertia and damping are control parameters, which can be altered in real-time, for an improved response to frequency disturbances. As shown in subsection 7.1.1, the increase in the inertia constant can help to reduce the ROCOF and to improve the frequency nadir in low-inertia power systems. The effect of the damping coefficient resembles the role of the active power droop in a conventional frequency control system. The increase in the damping coefficient can improve the frequency nadir. In this work, a new adaptive control method is proposed for altering these two parameters in Eq. (7.5), with the aim of reducing the ROCOF and improving the frequency nadir in low-inertia power systems. Details of the proposed adaptive inertia and damping method are discussed in subsection 7.2.2.

It is also important to ensure that the FESS is always capable of providing and absorbing active power when a frequency disturbance occurs. Therefore, the FESS is also equipped with a State of Charge (SOC) controller, which attempts to maintain the state of charge of 75 %, after the frequency deviation has passed. The choice of 75% is based on the fact that flywheels are not able to provide the nominal power at low speeds, as previously discussed. It was demonstrated in subsection 6.3.1 that the nominal power can be delivered only above the SOC of charge of approximately 25 %. Therefore, 75 % is the center of the range, in which the FESS can provide its full power. The state of charge of 75 % corresponds to 86.6 % of the maximum speed of the FESS. Therefore, ω_{ref} is set to $0.866\omega_{\max}$ in Figure 7.5. The state of charge controller uses a conventional PI controller in combination with a ramp limiter, to limit the effect of charging or discharging of the FESS on the grid, and it is activated only when the frequency has recovered fully to the nominal value.

In the block diagram of Figure 7.5, a frequency deadband ($\pm\omega$db) and a secondary frequency control block are included to provide a comprehensive frequency control framework. However, they are disabled in this work to show more clearly the effect of the inertia constant and the damping coefficient on the frequency response.

7.2.2 Adaptive Inertia and Damping for an Improved Response

As mentioned before, an adaptive strategy can be used for the inertia constant and the damping coefficient to improve the response of the FESS to frequency deviations. In recent years, numerous techniques and solutions have been proposed in literature on how these parameters should be altered during a frequency disturbance. The proposed methods in literature can be categorized into two main groups: methods based on online optimization algorithms [265], [296]–[300], and interval-based controllers [268]–[271]. In this thesis, a new interval-based controller is proposed, and its performance is compared with several previously reported methods in this category. It is important to note that none of the previously reported adaptive methods are suggested specifically for a FESS. However, they can be applied to any converter-interfaced system. Therefore, they are used as reference points in this work for a comparison with the proposed method.

In interval-based controllers, the time period of the frequency deviation is divided into several fragments, according to the sign of $\Delta\bar{\omega}_{\text{g}}\frac{d\Delta\bar{\omega}_{\text{g}}}{dt}$, as shown in Figure 7.6. When $\frac{d\Delta\bar{\omega}_{\text{g}}}{dt}$ and $\Delta\bar{\omega}_{\text{g}}$ have the same signs, i.e., $\Delta\bar{\omega}_{\text{g}}\frac{d\Delta\bar{\omega}_{\text{g}}}{dt} > 0$, the period is called the accelerating phase. This includes the first interval after a disturbance occurs until reaching the frequency nadir. When they have opposite signs or $\Delta\bar{\omega}_{\text{g}}\frac{d\Delta\bar{\omega}_{\text{g}}}{dt} \leq 0$, the interval is named the decelerating phase. In Figure 7.6, the accelerating phase is marked in red while the decelerating is colored in yellow. The naming of these periods resembles the behavior of a synchronous generators on its power-angle curve following a disturbance [268].

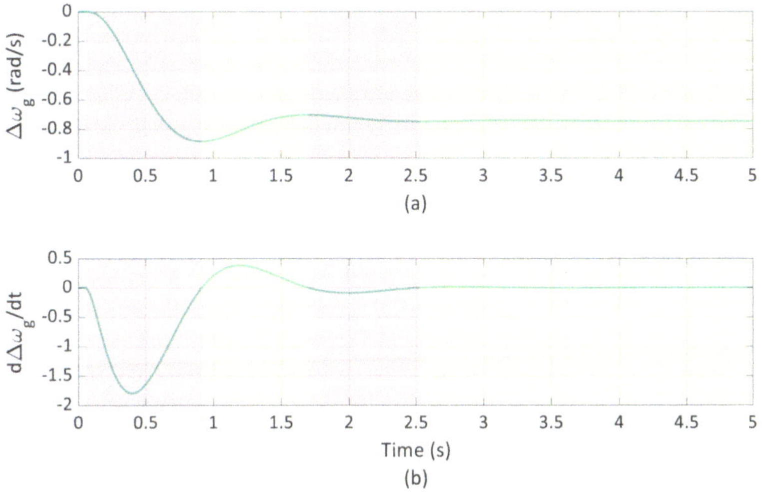

Figure 7.6: Segmentation of the frequency deviation period into the accelerating phase (red) and the decelerating phase (yellow), according to the sign of $\Delta\bar{\omega}_{\mathrm{g}}\frac{d\Delta\bar{\omega}_{\mathrm{g}}}{dt}$.

As shown in subsection 7.1.1, having a large value for the inertia constant leads to a reduction of the maximum ROCOF observed during a frequency disturbance. However, maintaining the high inertia value leads to a slow recovery of the frequency, and an increased setting time for the frequency [268]. To solve the problem of the slow frequency recovery, a bang-bang control strategy for the inertia value is proposed in [268], where the inertia changes between two discrete values, depending on being in the accelerating or decelerating phases. Using this approach, when the frequency is moving away from the nominal value (accelerating phase), a high inertia is applied to resist its changes. As soon as the frequency begins to recover towards a new steady-state value (decelerating phase), the emulated inertia is set to zero to speed up the frequency recovery. However, the second degree of freedom, the damping coefficient is neglected in [268]. Therefore, in [269], a similar approach using the bang-bang control has been applied on both the damping and the inertia. The use of the damping in this design improves the frequency nadir and the frequency settling time. In both control designs using the bang-bang approach, the inertia constant and the damping coefficient can change only between two predetermined discrete values, which are independent from the severity of the frequency disturbance. In the adaptive inertia controller in [271], however, an adaptive component is added to the inertia, that scales with the amplitude of $\Delta\bar{\omega}_{\mathrm{g}}\frac{d\Delta\bar{\omega}_{\mathrm{g}}}{dt}$. But this work, again, neglects the use of the damping coefficient. A self-adaptive controller is proposed in [270], which considers both the damping and the inertia constant. However, the damping is increased only after the frequency nadir (decelerating phase) to damp out the subsequent frequency oscillations. Therefore, it cannot influence the frequency nadir at all. The main issue with the proposed adaptive controllers in [271] and [270] is that the changes in the inertia and damping are not as fast as the bang-bang controllers, where the changes are almost immediate. In low-inertia power systems, where a high ROCOF is observed within the first instances of the frequency disturbance, it is important to reach a high inertia constant as fast as possible to effectively limit the maximum ROCOF. Enabling a high inertia value later than necessary can have little to no impact on the maximum ROCOF observed during the disturbance. A summary and a comparison of the interval-based adaptive controllers studied in this work is presented in Table 7.3.

Table 7.3: A comparison of the proposed inertia emulation control with several interval-based methods in literature.

Literature	Bang-bang control	Self-adaptive control	Considering damping	Considering limits of the active power source
[268]	Yes	No	No	No
[269]	Yes	No	Yes	No
[271]	No	Yes	No	No
[270]	No	Yes	Yes*	No
Proposed	Yes	Yes	Yes	Yes

* The higher damping is applied only after the frequency nadir.

In this work, we combine the advantages of bang-bang control approaches, which guarantee a rapid change of the control parameters, with self-adaptive controllers, that ensure a more proportional response to the frequency disturbance. Moreover, we take advantage of the damping coefficient to help reducing the frequency nadir. The proposed adaptive control for the inertia constant and damping coefficient is described by Eq. (7.7) and (7.8), respectively.

$$H = \begin{cases} H_1 + K_H \left|\frac{d\Delta\bar{\omega}_g}{dt}\right| & \text{, if } (\Delta\bar{\omega}_g \frac{d\Delta\bar{\omega}_g}{dt} > 0) \cap \left(\left|\frac{d\Delta\bar{\omega}_g}{dt}\right| > \varepsilon_H\right) \\ H_2 & \text{, if } (\Delta\bar{\omega}_g \frac{d\Delta\bar{\omega}_g}{dt} \leq 0) \cup \left(\left|\frac{d\Delta\bar{\omega}_g}{dt}\right| \leq \varepsilon_H\right) \end{cases}. \tag{7.7}$$

$$D = \begin{cases} D_1 + K_D \left|\Delta\bar{\omega}_g\right| & \text{, if } (\Delta\bar{\omega}_g \frac{d\Delta\bar{\omega}_g}{dt} > 0) \cap \left(\left|\frac{d\Delta\bar{\omega}_g}{dt}\right| > \varepsilon_D\right) \\ D_2 + K_D \left|\Delta\bar{\omega}_g\right| & \text{, if } (\Delta\bar{\omega}_g \frac{d\Delta\bar{\omega}_g}{dt} \leq 0) \cup \left(\left|\frac{d\Delta\bar{\omega}_g}{dt}\right| \leq \varepsilon_D\right) \end{cases}. \tag{7.8}$$

As seen from Eq. (7.7), as soon the frequency disturbance occurs (accelerating phase), the inertia constant is instantaneously increased from a lower value (H_2) to a higher value (H_1), which is further increased by a proportional term ($K_H \left|\frac{d\Delta\bar{\omega}_g}{dt}\right|$), that scales linearly with the absolute value of the ROCOF. Therefore, using the proposed approach, the inertia constant increases rapidly using the bang-bang approach, and according to the severity of the frequency disturbance, an even higher inertia constant is applied. When the frequency begins to recover (decelerating phase), the inertia constant is reduced immediately to a value near zero (H_2) to avoid the increase in frequency settling time. The condition to apply the higher inertia is being in the accelerating phase and not being in steady-state conditions, which is determined by $\left|\frac{d\Delta\bar{\omega}_g}{dt}\right| > \varepsilon_H$. The addition of the second condition, avoids unnecessary switching between the two states, which are observed in [268], where ε_H is set to zero. A low inertia value is also applied when the frequency has reached a steady-state condition, i.e., $\left|\frac{d\Delta\bar{\omega}_g}{dt}\right| \leq \varepsilon_H$.

A similar controller design is suggested for the damping coefficient. However, the adaptive component of the damping coefficient ($K_D \Delta\bar{\omega}_g$) scales with the frequency error, rather than the ROCOF. Also, the adaptive term is maintained when the frequency is recovering, while the constant term is reduced from D_1 to D_2. This helps improving the damping of the subsequent frequency oscillations after the nadir, while the reduction of the constant term helps to limit the energy drawn from the FESS.

The differences between the proposed controller and the reviewed adaptive methods in literature is summarized in Table 7.3. As seen in Table 7.3, the proposed controller also considers the limitations of the active power source used for inertia emulation, which is discussed in the next subsection.

7.2.3 Parameters for the Inertia Emulation Control using a FESS

Many related works on adaptive inertia emulation techniques do not consider the limitations of the active power source behind the converter, and assume an ideal DC-link with unlimited energy resources, as in [273]–[276]. This section proposes a simple methodology to include the state of charge and the practical limitations of the FESS at low rotational speeds into the control design by adjusting the parameters of the proposed inertia emulation controller according to the state of charge of the FESS.

In the proposed control design for the inertia constant, shown in Eq. (7.7), H_1 and K_H are the control parameters that determine the emulated inertia, which are used to calculate the active power injection according to the ROCOF. When the FESS is fully charged at its maximum speed of ω_{max}, these two parameters should be at their maximum level to provide the maximum inertia. However, it is proposed in this work that these parameters are scaled down with the reduction of the state of charge of the FESS. Using this approach, the FESS will provide inertia support proportional to its state of charge.

Therefore, the inertia constant H_1 is equal to its maximum value for a fully charged FESS, $\mathrm{H}_{1\mathrm{max}}$, and for lower values of state of charge,

$$H_1 = \frac{\frac{1}{2}\mathrm{J_f}\omega_{\mathrm{m}}^2}{\frac{1}{2}\mathrm{J_f}\omega_{\mathrm{max}}^2}\mathrm{H}_{1\mathrm{max}} = \left(\frac{\omega_{\mathrm{m}}}{\omega_{\mathrm{max}}}\right)^2\mathrm{H}_{1\mathrm{max}} = \mathrm{SOC}\cdot\mathrm{H}_{1\mathrm{max}}. \tag{7.9}$$

Figure 7.7(a) shows the inertia constant H_1 as a function of the speed of the FESS according to Eq. (7.9), and for a maximum inertia constant of 5.9 s. A similar function to Eq. (7.9) is used for K_H with the maximum value of K_{Hmax}, i.e., $K_H = \mathrm{SOC}\cdot K_{\mathrm{Hmax}}$. The parameter H_2 is constant and close to zero to speed up the frequency recovery.

Furthermore, we integrate the power limitations of the FESS at low rotational speed into the controller. This limitation is integrated into the damping coefficient D_2, as it determines the FESS contribution in steady-state conditions. The damping coefficient D_2 is reduced from its maximum value $D_{2\mathrm{max}}$ when the FESS reaches the low speed threshold, according to the speed of the FESS, as shown in Figure 7.7(b). A similar function is used for D_1 and K_D, with the maximum values of $D_{1\mathrm{max}}$ and K_{Dmax}, respectively, as shown in Figure 7.7.

Using the proposed approach, the FESS contribution to frequency deviations is reduced automatically, when the FESS reaches lower speeds and reaches zero, when the FESS is emptied. This is a mandatory requirement, when using a real FESS, where the physical system constrains must be considered.

As the damping coefficient D_2 determines the steady-state contribution of the FESS, it has the same effect as the droop coefficient. Therefore, we chose the maximum value of D_2 to be 40 p.u., which corresponds to a droop value of 2.5 %. The value for D_1 is chosen to be slightly higher to help to improve the frequency nadir, as discussed earlier. The choice of K_{Hmax} and K_{Dmax} determines the effect of the instantaneous ROCOF and frequency on the inertia constant and the damping coefficient, respectively. These are design parameters, which can be selected according the maximum ROCOF and frequency deviation expected in a system. The value of ε_{H} and ε_{D}, which determine the transition condition in the proposed controller, should be higher than the noise level expected on the ROCOF and frequency estimation, in order to avoid unwanted triggering between the two control states. In this respect, using a differential operator to calculate the ROCOF can lead to noise amplification and inaccurate triggering of the controller. Therefore, in the next section, a method is described to calculate the ROCOF without using this operator. The parameters used for the proposed inertia emulation control in this work are given in Table 7.4.

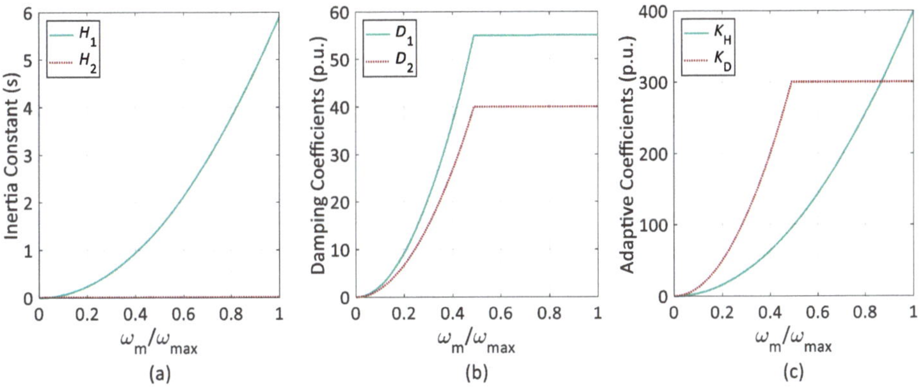

Figure 7.7: The variations of the inertia emulation control parameters with respect to the speed of the FESS. (a) The inertia constants, H_1 and H_2, (b) the damping coefficients D_1 and D_2, and (c) the adaptive coefficients K_H and K_D.

Table 7.4: The parameters used for the proposed inertia emulation controller.

Parameter	**Value**	**Parameter**	**Value**
H_{1max}	5.9 s	D_{1max}	55 p.u.
H_2	0.01 s	D_{2max}	40 p.u.
K_{Hmax}	400 p.u.	K_{Dmax}	400 p.u.
ε_H	0.005 p.u.	ε_D	0.005 p.u.

7.2.4 Estimation of the Frequency and its Derivative

Inertia emulation techniques require estimated values of the frequency and the frequency derivative (ROCOF). The Synchronous Reference Frame Phase-locked Loop (SRF-PLL) has been used often in inertia emulation controllers in literature [290], [294], [295], [300], as the conventional method for estimating the frequency. However, it has been reported that the use of the SRF-PLL in inertia emulation applications can lead to instabilities and noise amplification associated with the differential operation, used for calculating the ROCOF [301]–[303]. Furthermore, the SRF-PLL performs poorly under unbalanced and distorted voltage conditions, often observed in weak grids, such as microgrids [304]. Although the performance of the SRF-PLL in such conditions can be improved by additional filters or moving window functions, these methods introduce more delays into the frequency estimation. For the application of inertia emulation in low-inertia power systems, a fast detection of frequency changes is required, and such delays can impair the benefits of inertia emulation.

Therefore, in this work, it is suggested to use the Dual Second-Order Generalized Integrator Frequency-Locked Loop (DSOGI-FLL), introduced in [304]. Using the DSOGI-FLL has the significant advantage that it can estimate the frequency derivative without using the differential operator, avoiding the problems of noise amplification and unrealistic outputs. The DSOGI-FLL only uses integrators and multiplication of filtered signals to estimate the ROCOF. The superior performance of the DSOGI-FLL over the SRF-PLL for the estimation of the ROCOF has been demonstrated by simulations and experiments in [302]. The same work shows that the common stability issues of the SRF-PLL are not observed in the DSOGI-FLL. In addition, the DSOGI-FLL is based on adaptive filters. Therefore, it performs well under distorted, faulty, and unbalanced voltage conditions [304]. The block diagram of the DSOGI-FLL is depicted in Figure 7.8.

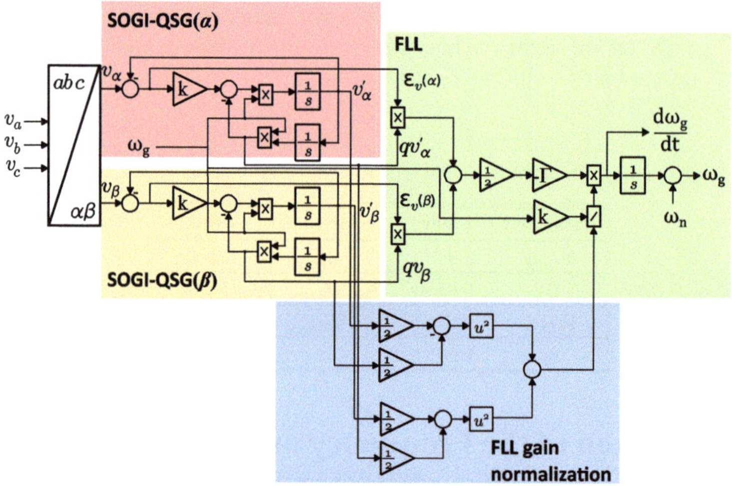

Figure 7.8: The block diagram of the Dual Second-Order Generalized Integrator Frequency-Locked Loop (DSOGI-FLL) (adopted from [304]).

As seen in Figure 7.8, the DSOGI-FLL consists of two Second-order Generalized Integrator Quadrature Signal Generators (SOGI-QSG) and a Frequency-locked Loop (FLL), with a gain normalization block. The two SOGI-QSG receive the measured voltage in the stationary reference frame (v_α and v_β), and generate filtered in-phase (v'_α and v'_β) and in-quadrature (qv'_α and qv'_β) signals of the input voltages with the same amplitude. The transfer function of these conversions for the in-phase ($G^d_{\alpha\beta}(s)$) and for the in-quadrature ($G^q_{\alpha\beta}(s)$) components are shown in Eq. (7.10) and (7.11), respectively, which are identical for both α and β components.

$$G^d_{\alpha\beta}(s) = \frac{v'_{\alpha\beta}}{v_{\alpha\beta}} = \frac{\mathrm{k}\omega_{\mathrm{g}} s}{s^2+\mathrm{k}\omega_{\mathrm{g}} s+\omega_{\mathrm{g}}^2}. \tag{7.10}$$

$$G^q_{\alpha\beta}(s) = \frac{qv'_{\alpha\beta}}{v_{\alpha\beta}} = \frac{\mathrm{k}\omega_{\mathrm{g}}^2}{s^2+\mathrm{k}\omega_{\mathrm{g}} s+\omega_{\mathrm{g}}^2}. \tag{7.11}$$

In these equations, ω_{g} is the estimated frequency by the FLL and k is the SOGI-QSG gain, which is a design parameter. By looking at Eq. (7.10) and (7.11), it can be seen that the SOGI-QSG control loops are basically a band-pass filter for the in-phase components and a low-pass filter for the in-quadrature signals. To have the optimum damping ratio (ξ) of 0.707, the SOGI-QSG gain k is set to $\sqrt{2}$, as recommended in [304]. This choice leads to an optimum balance between a fast dynamic response and the filtering capability of the SOGI-QSG. The FLL uses the error in calculating the in-phase components and the filtered in-quadrature components to calculate the frequency derivative as

$$\frac{d\bar{\omega}_{\mathrm{g}}}{dt} = -\frac{\Gamma}{2}((v'_{\alpha} - v_{\alpha})qv'_{\alpha} + (v'_{\beta} - v_{\beta})qv'_{\beta}). \tag{7.12}$$

In Eq. (7.12), Γ is the FLL gain, which is assumed to be equal to 100, as recommended in [251]. This choice of Γ leads to a relatively fast frequency setting time of the frequency measurement. It is clear from Eq. (7.12) that no differential operation is involved in estimating the ROCOF, when using the DSOGI-FLL. The gain normalization block ensures that FLL receive signals with unity magnitude, avoiding errora in frequency estimation [251]. The frequency itself is estimated by integrating over the calculated frequency derivative.

7.3 Simulation Results of the Adaptive Inertia Emulation using a high-speed FESS

The performance of the proposed adaptive inertia emulation controller for the high-speed FESS is initially evaluated using offline simulations in MATLAB/Simulink environment. Two difference scenarios in a low voltage microgrid are presented to show the effectiveness of the proposed method in improving the frequency dynamics during power imbalances. For a comparison with the current state-of-the-art, three of the interval-based adaptive inertia controllers, previously reported in the literature, are also simulated in each scenario.

The simulation scenarios are carried out in a low voltage microgrid, based on the CIGRE European low voltage network benchmark [123], and its variation in [305]. Among the available benchmark grids with openly available data, this network benchmark has the closest characteristics to a typical German low voltage distribution grid. The high-speed FESS together with two PV systems and a diesel-based Combined Heat and Power (CHP) system are added to the CIGRE benchmark to form a microgrid and allow an autonomous operation of the system. The single-line diagram of the microgrid is shown in Figure 7.9. The FESS is installed near the transformer to provide frequency and voltage support during islanding, while the PV systems and the CHP are located near the loads. The microgrid is capable of being disconnected from the main supply and operate in the autonomous mode, when necessary. By a power imbalance in the autonomous operation of the microgrid, the frequency can deviate greatly, as the synchronous generation of the CHP is the only source of traditional inertia in the system. During an imbalance of generation and demand, the CHP slowly covers the demand, while the FESS has the task of providing a faster response and additional inertia support during the first seconds of the frequency disturbance. The PV systems are modelled according to the generic two-stage model presented in [261] and are being operated at their maximum power point. Therefore, they are not able to take part in regulating the frequency during under-frequency events. The parameters of the diesel-based CHP and the PV systems are given in the Appendix B. The FESS parameters are the same as the ones use in Chapter 5 and 6, and the inertia emulation parameters are given in Table 7.4.

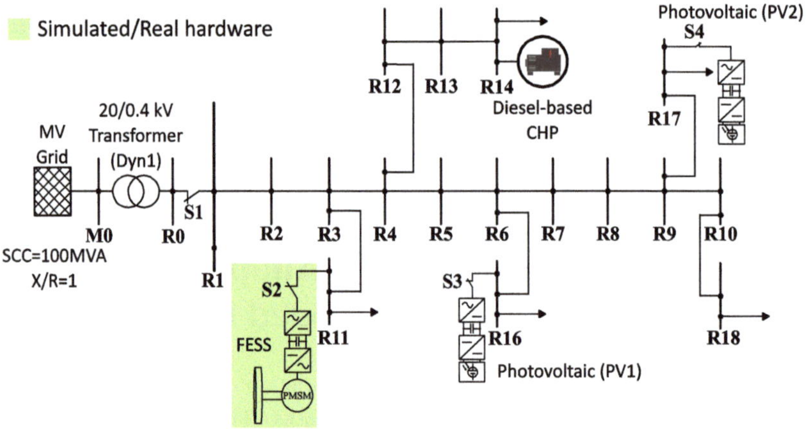

Figure 7.9: The single-line diagram of the test microgrid based on the CIGRE European low voltage network benchmark.

Two different scenarios are simulated, which are the microgrid islanding and the intermittent generation of the PV systems in the autonomous mode of operation. In each scenario, six difference cases are simulated:

- **Case 1:** No FESS.
- **Case 2:** FESS with a conventional droop control and no inertia emulation.
- **Case 3:** FESS with inertia emulation control and the bang-bang control for both the inertia and the damping, as in [269].
- **Case 4:** FESS with inertia emulation control and adaptive inertia control, as in [271].
- **Case 5:** FESS with inertia emulation control and adaptive inertia and damping control, as in [270].
- **Case 6:** FESS with inertia emulation control and the proposed adaptive inertia and damping method, as described in section 7.2.2.

A summary of the characteristics of the controllers in all cases is presented in Table 7.5. For a fair comparison, the same maximum inertia constant and damping of Table 7.4 is used for the bang-bang control in case 3. For the controllers in case 4 and 5, where a single value of inertia constant is used, the average of inertia and damping values of Table 7.4 is used. The simulation results in the two scenarios and for all the six cases are provided next.

Table 7.5: A summary of the controller characteristics in all cases.

Case	FESS	Controller type	Adaptive control method
Case 1	No	N/A	N/A
Case 2	Yes	Droop	N/A
Case 3	Yes	Inertia Emulation	[269]
Case 4	Yes	Inertia Emulation	[271]
Case 5	Yes	Inertia Emulation	[270]
Case 6	Yes	Inertia Emulation	Proposed.

7.3.1 Scenario A: Microgrid Islanding

In this scenario, the microgrid is initially drawing 80 kW of power from the medium voltage grid, while the PV systems and the CHP system are only partially supporting the loads. The FESS is charged to reach the state of charge of 75 % using the state of charge controller, prior to the disturbance. The microgrid is decoupled instantaneously from the main supply by opening the circuit breaker S1, and it undergoes into the islanded mode of operation. The power deficiency in the microgrid leads to a sudden drop in the microgrid frequency. The FESS attempts to limit the ROCOF and frequency nadir, while the CHP system slowly increases its output to compensate for the power deficiency.

The simulation results for all the six cases in this scenario are presented in Figures 7.10 and 7.11. It can be seen from Figures 7.10(a) and (b), that the use of the FESS with the proposed adaptive inertia emulation controller, presented as case 6, leads to the highest reduction in the maximum ROCOF and the most improved frequency nadir. As shown in Figure 7.10(b), the FESS with the proposed controller reduces the maximum ROCOF by 28 %, from 1.72 Hz/s in case 1, to 1.24 Hz/s in case 6. Furthermore, it performs better than the previously proposed adaptive inertia controllers, presented as case 3 to 5, which reduce the maximum ROCOF to 1.34 Hz/s at best. The superior performance of the proposed design in reducing the ROCOF is due to quickly reaching a higher inertia constant, as shown in Figure 7.10(c), using a combination of the bang-bang control approaches and the self-adaptive ones. In case 3, where only the bang-bang control is used, the inertia constant also changes very rapidly. However, it only reaches a predetermined constant value ($H_{1\max}$), independent from the ROCOF, and remains constant until the frequency nadir is reached, which limits its effectiveness. The adaptive inertia controller implemented in case 4 changes the inertia constant slowly, and therefore, it performs worse than the proposed controller. In the self-adaptive controller of case 5, the inertia constant is only increased by the instantaneous value of the ROCOF. Therefore, the increase is not as high and as fast as in the proposed design. As a result of combing the bang-bang control and the self-adaptive approach in the proposed design, the inertia constant immediately reaches the high value of 4.42 s (H_1 at SOC of 75%), which is increased even further by the instantaneous value of the ROCOF, reaching an inertia constant of 7.8 s quickly after islanding. Among the previously reported control designs, case 3 and 5 show a similar behavior in reducing the maximum ROCOF. The controller in case 5 reduces the ROCOF slightly better than the one in case 3, due the fact that the initial ROCOF in this scenario is high enough to reach a higher inertia constant than the pre-given value in case 3. Although a higher inertia constant is reached in case 4, in comparison to case 3, the increase is not fast enough to effectively limit the maximum ROCOF. Thus, the worst performance is observed in case 4 among the cases 3 to 5. The maximum ROCOF and frequency nadir in all cases are summarized in Table 7.6.

The proposed controller also results in the largest improvements in the frequency nadir, by taking advantage of the second degree of the freedom, the damping coefficient. As seen in Figure 7.10(a), the use of the FESS with a conventional droop control, presented as case 2, improves the frequency nadir from 48.84 Hz, to 49.2 Hz. However, with a proper use of the damping coefficient in the proposed design, the frequency nadir can be improved even further, to reach 49.35 Hz. This is a 44 % reduction in the maximum frequency deviation in comparison to case 1, and 18.7 % in comparison to case 2. This is achieved by increasing the damping coefficient immediately after the islanding, and further increasing it according to the instantaneous changes in the frequency, as shown in Figure 7.10(d). Between the studied cases, the controller in case 4 does not use the damping coefficient, leading do the worst frequency nadir. Following case 6, the most improved frequency nadir can be observed in case 3, as unlike the controller in case 5, a higher damping coefficient is applied before reaching the frequency nadir. It should also be noted, that the improved frequency nadir is also a side effect of the reduction in the ROCOF in the proposed design using the inertia constant. Using the FESS with the proposed control design also shows a better damping of frequency oscillations after the frequency nadir, as seen in Figure 7.10(a). This is due to maintaining the adaptive component of the damping coefficient, after reaching the frequency nadir.

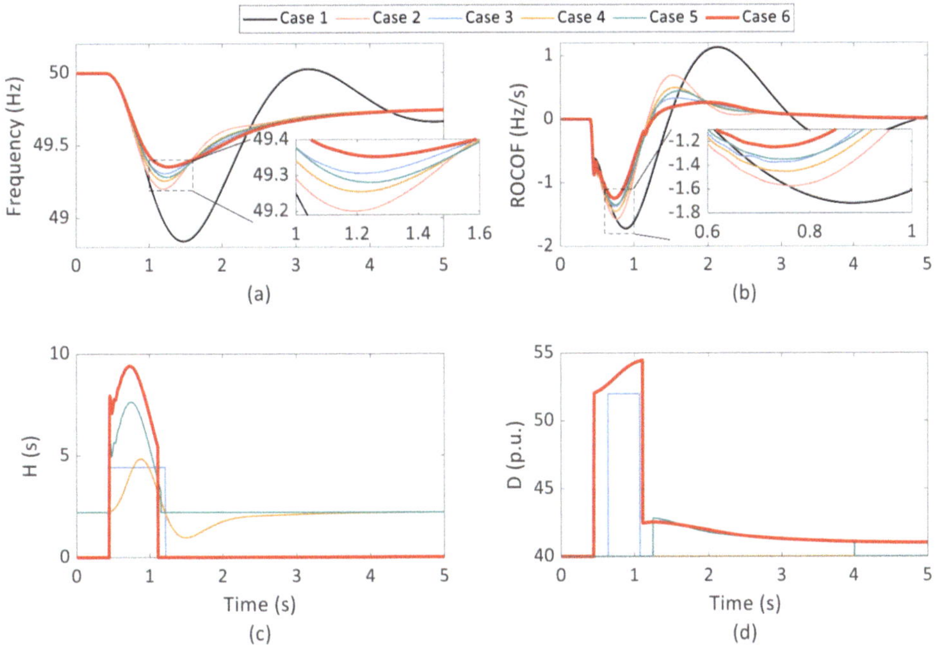

Figure 7.10: Simulation results for scenario A: microgrid islanding. (a) Frequency, (b) ROCOF, (c) Inertia constant (H), and (d) Damping coefficient (D).

Figure 7.11 shows the active power of the FESS and its state of charge variations during the first five seconds of the islanding scenario. As seen in Figure 7.11(a), the use of the proposed controller leads to a more prompt and faster response of the FESS, which leads to the improved grid frequency dynamics during the islanding. The controller design in case 3 results in a slightly higher peak in power than the proposed design; however, a better performance in terms of the frequency is not observed, as the increase in power is slower than in the proposed control design. In case 3 and 6, where the bang-bang control and a higher damping coefficient from the beginning of the frequency incident are applied, a higher peak in the FESS power is observed. However, the power is reduced significantly using the bang-bang approach, after reaching the frequency nadir. Therefore, the total energy drawn from the FESS in all cases do not differ significantly, as seen in Figure 7.11(b). The proposed control design increases the energy drawn from the FESS by only 0.15 %, in comparison to case 2, and by 0.07 %, in comparison to case 3 and 5. It can be concluded, that the proposed adaptive inertia emulation controller leads to an improved frequency dynamic of the microgrid, in terms of reducing the maximum ROCOF and improving the frequency nadir, while not using significantly more energy from the FESS. This is an advantage in a long-term autonomous operation of the microgrid with continuous variations in load and generation.

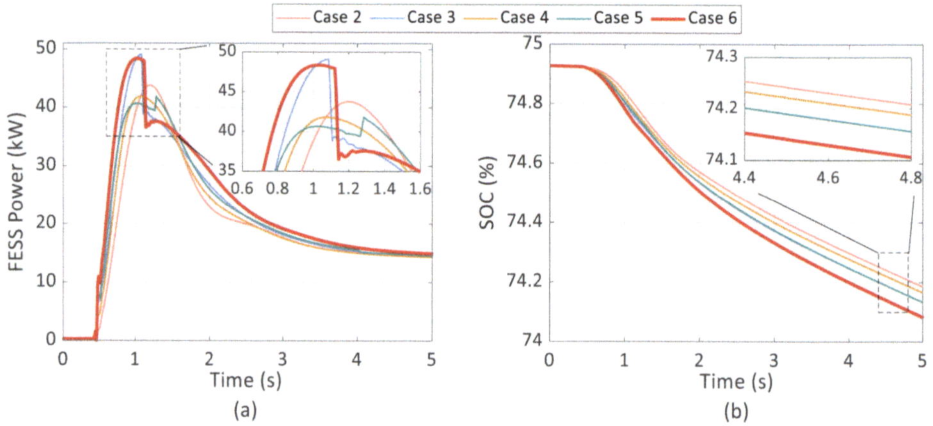

Figure 7.11: Simulation results for scenario A: microgrid islanding. (a) FESS Power, and (b) FESS State of Charge (SOC).

Table 7.6: The maximum ROCOF and frequency nadir for all the simulated scenarios and cases.

	Frequency Nadir (Hz)		Maximum ROCOF (Hz/s)	
Case	**Scenario A**	**Scenario B**	**Scenario A**	**Scenario B**
Case 1	48.84	49.04	1.72	1.15
Case 2	49.20	49.22	1.57	1.06
Case 3	49.31	49.29	1.37	0.89
Case 4	49.25	49.26	1.45	0.98
Case 5	49.28	49.27	1.34	0.95
Case 6	49.35	49.33	1.24	0.82

7.3.2 Scenario B: Drop in the PV Generation

In scenario B, the microgrid is disconnected from the main supply, after an islanding scenario, similar to the one in scenario A. The CHP system and the two PV systems are supporting the loads, while the FESS is injecting a small amount of power according to the value of D_2 and the steady-state frequency error. A sudden sharp drop in the solar irradiance density is simulated, reducing it from 1000 W/m^2 to 150 W/m^2, which could be caused by passing clouds. This results in a drastic drop of the PV generation in the microgrid, from 70 kW to 10.5 kW. The power deficit leads to a drop in the frequency of the microgrid. The FESS increases rapidly its output to limit the ROCOF and frequency nadir, while the CHP system slowly covers the lack of generation by the PV systems.

The simulation results for the six cases in this scenario are shown in Figures 7.12 and 7.13. Similar to scenario A, the use of the FESS with the proposed controller leads to the highest reduction in the maximum ROCOF and frequency nadir during this frequency disturbance. It reduces the maximum ROCOF from 1.15 Hz/s in case 1, to 0.82 Hz/s in case 6. It can be seen from Figure 7.12(c) that this is achieved by quickly reaching a higher inertia constant in the proposed controller in comparison to the other cases. In contrast to the scenario A, the controller in case 3 performs better than case 5 in this scenario. This is due to the fact, that the inertia constant in case 5 increases only with the ROCOF, and since the ROCOF in this scenario is lower than the one in scenario A, case 3 with a predetermined increase in the inertia constant shows a better performance. Despite eventually reaching a higher inertia constant in case 5, the increase is not fast enough to limit the maximum ROCOF.

Similarly, the highest improvement in the frequency nadir is seen using the proposed control design, where the frequency nadir reaches only 49.34 Hz, in comparison to 49.04 Hz in case 1, and 49.22 Hz in case 2. Following case 6, case 3 shows the best performance in terms of the frequency nadir, by reaching 49.29 Hz. This is due to the fact that it also uses a higher damping coefficient before reaching the frequency nadir.

Figures 7.13(a) and (b) show the active power and the state of charge of the FESS, respectively. As seen in Figure 7.13(a), the FESS is injecting about 12 kW of power, prior to the disturbance. Similar to the previous scenario, with the proposed controller, the FESS increases its power output faster than in the other cases. It also reaches a higher peak in power in this scenario, but due to the reduction of the power after reaching the frequency nadir, there is only a slight increase in the energy required from the FESS, when using the proposed controller. This can be seen by comparing the state of charge before the disturbance with the one after the five seconds into the disturbance. It is important to note that the state of charge values in different cases are different prior to the disturbance, as the drop in the PV generation is simulated following an islanding scenario using different controllers.

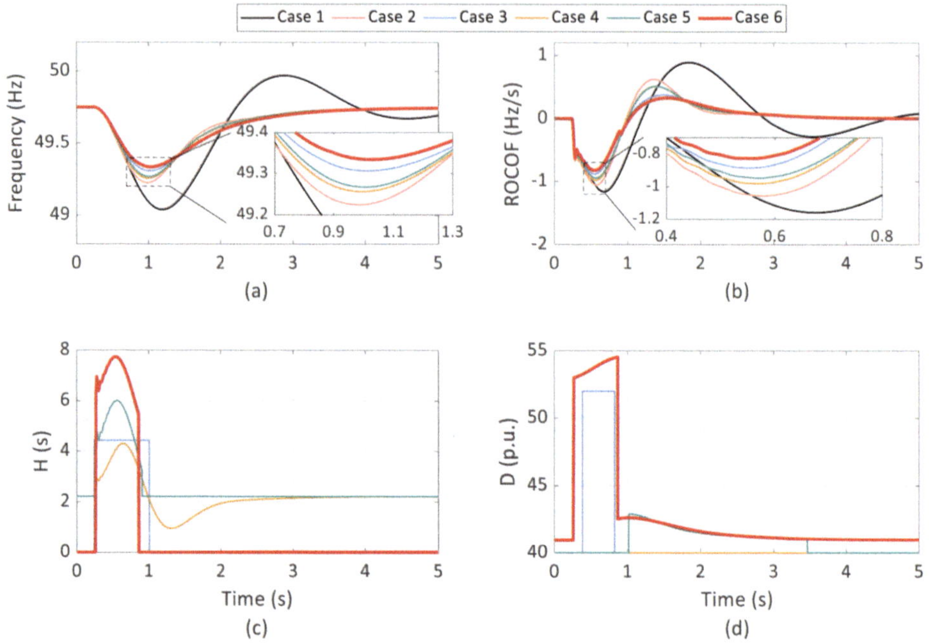

Figure 7.12: Simulation results for scenario B: drop in PV generation. (a) Frequency, (b) ROCOF, (c) Inertia constant (H), and (d) Damping coefficient (D).

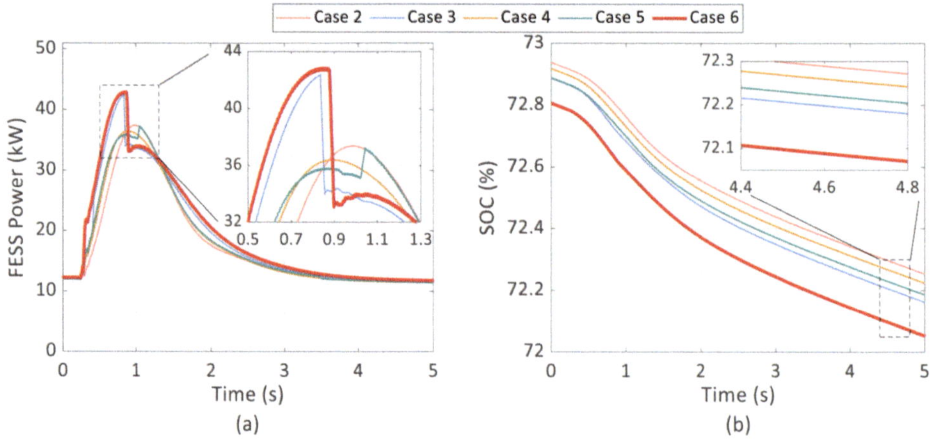

Figure 7.13: Simulation results for scenario B: drop in PV generation. (a) FESS Power, and (b) FESS State of Charge (SOC).

7.4 PHIL Testing of a High-speed FESS for Inertia Emulation

To experimentally validate the proposed inertia emulation control for the FESS, the controller has been implemented on the 60 kW high-speed FESS, using the concept of rapid control prototyping. The performance of the FESS using the proposed adaptive inertia emulation controller has been evaluated using PHIL testing. The two simulated scenarios, i.e., the microgrid islanding and the drop in PV generation, have been repeated in real-time for conducting the PHIL testing of the FESS. In each scenario, all six cases have been investigated, leading to 12 experiments in total.

7.4.1 Real-time Simulation of the Microgrid

For conducting the PHIL testing, a real-time simulation of the grid under study is mandatory. The CIGRE-based low voltage microgrid shown in Figure 7.9 is simulated in real-time on the Opal-RT's OP5700 real-time simulator with simulation time steps of only 24 µs. This is achieved by using the State-space Nodal (SSN) solver [109] to decouple the model at bus R4, creating three groups of equations that can be solved almost independently (see subsection 3.1 for more information), and neglecting the line capacitances due to their short lengths [117]. The relatively low simulation time step leads to a low total loop delay, which is a main contributor to the PHIL stability, as discussed in subsection 3.2.2.

7.4.2 Implementing the Inertia Emulation Controller on the real FESS

The same PHIL setup and commercial high-speed FESS, which were described in Chapter 6, are also used for the experiments in this chapter.

To implement the proposed inertia emulation controller on the commercial FESS, the controller is also simulated in real-time on the OP5700 real-time simulator, as an external controller. It measures the frequency and frequency derivative from the grid voltage at bus R11, and generates the active power reference for the high-speed FESS. This concept, where the controller is simulated in real-time, while it is being tested on a real hardware is known as rapid control prototyping. The active power reference generated from the simulated controller is sent in real-time to the FESS internal controller through a 4-20 mA analog channel. In this work, the analog connection has been preferred to the standard Modbus communication, as the Modbus communication in known to be slow and nondeterministic in signal transmission, which is not suitable for this application. A schematic of the described signal connections and the PHIL testing setup is shown in Figure 7.14.

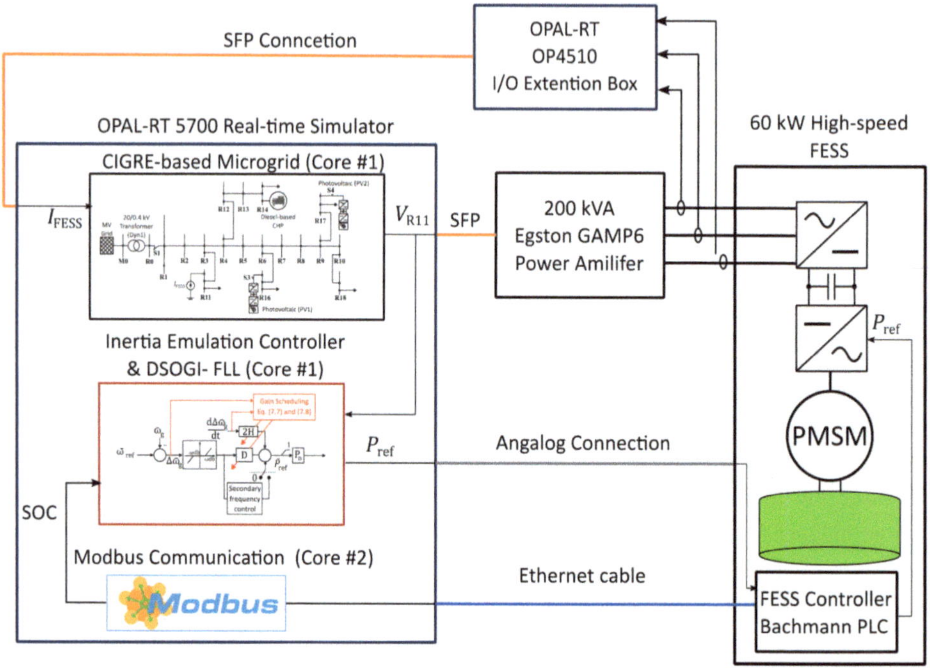

Figure 7.14: The signal connections and PHIL setup for implementing the inertia emulation controller on the FESS using the concept of rapid control prototyping.

In Figure 7.14, V_{R11} is the three-phase grid voltage at bus R11, where the FESS is virtually connected. This voltage is sent via a dedicated SFP connection to the Egston 200 kVA power amplifier, which is feeding the FESS. It is also sent within the simulator to the DSOGI-FLL and the inertia emulation controller. The DOSGI-FLL receives this voltage and estimates the frequency and frequency derivative, as described in subsection 7.2.4. The estimated frequency and ROCOF are sent to the simulated inertia emulation controller, where the active power reference (P_{ref}) is calculated. The inertia emulation controller also receives the state of charge of the FESS form the FESS controller via the Ethernet connection and the Modbus protocol. The active power reference is sent from the real-time simulator to the FESS controller via an analog connection. The FESS controller sends the active power reference to the MSC controller, where it generates the torque reference from the active power reference. The FESS injects or absorbs power, accordingly to the received set point. The FESS currents (I_{FESS}) are measured at the FESS terminals, sent to the real-time simulator via another SFP connection, and inserted into the simulated grid at bus R11. Similar to the PHIL tests in Chapter 6, the voltage-type ideal transformer method [256] is used for interfacing the FESS and the real-time simulation, and a first-order low-pass filter with a cut-off frequency of 100 kHz is applied on the current measurements fed back to the real-time simulation.

To illustrate the functionally of the setup, steps in the active power reference are sent from the real-time simulator to the FESS and the response of the FESS to the received set point is measured. The results are shown in Figure 7.15. First, a 60 kW step in the active power reference is sent to the FESS, while initially the FESS is injecting no power. It can be observed from Figure 7.15(a) that the FESS reaches the active power set point of 60 kW in just over 25 ms, which indicates a very quick response from FESS. Next, while the FESS is being charged at its nominal power, the active power set point is changed stepwise from -60 kW to +60 kW, resulting in a 120 kW step change in the active power. As shown in Figure 7.15(b), it takes less than 32 ms for the FESS to go from charging at full power to discharging at the same rate.

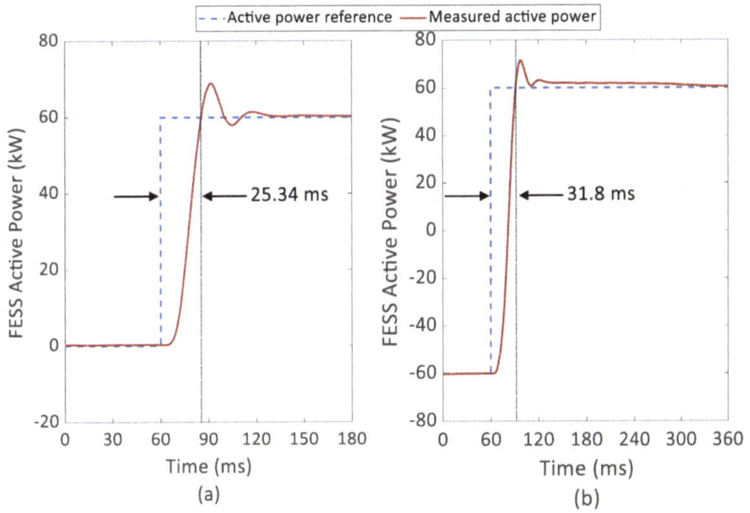

Figure 7.15: Response of the FESS to steps in the active power reference, sent from the OP5700 real-time simulator. (a) A 60 kW step, from steady-state operation and no power injection. (b) A 120 kW step, from charging at the nominal power to discharging at the same rate.

The same scenarios and cases, described in the simulations of section 7.3, are repeated using the real-time simulation of the microgrid and the real FESS. The obtained PHIL testing results are presented in the next subsections.

7.4.3 Scenario A: Microgrid Islanding

In this scenario, the simulated microgrid, running in real-time on the OP5700 real-time target, is decoupled from the main supply. The medium voltage grid is providing 80 kW of power to the microgrid, prior to the islanding. For all the six cases, before the islanding, the FESS is charged using the state of charge controller to reach the state of charge of 75 %, which corresponds to a rotational speed of approximately 39,000 rpm.

The results of the PHIL testing of the FESS for this scenario are shown in Figures 7.16 and 7.17. The frequency and the ROCOF of the microgrid during the islanding are depicted in Figure 7.16(a) and (b), respectively. It can be observed, that similar to the simulation results, the best performance in terms of reducing the maximum ROCOF and improving the frequency nadir is achieved using the proposed adaptive inertia emulation controller. The use of the high-speed FESS with the proposed inertia emulation controller, reduces the maximum ROCOF during the islanding scenario by 25 %, from 1.72 Hz/s in case 1, to 1.3 Hz/s in case 6. The superior performance of the proposed control design is achieved by rapidly reaching a higher inertia constant, as shown in Figure 7.16(c), by a combination of the bang-bang control methods and self-adaptive ones. The inclusion of the adaptive component in the inertia constant of the proposed controller leads to a better performance in comparison to the simple bang-bang control of case 3, where the inertia values in each interval are constant. Similarly, exploiting the bang-bang control approach in the proposed design results in a faster change of parameters in comparison to the self-adaptive controller of case 4.

Among the previously reported adaptive inertia controllers, case 3 and 5 show similar results, reducing the maximum ROCOF to only 1.4 Hz/s. Despite reaching a higher inertia constant in case 4, in comparison to case 3, a further improvement of the ROCOF is not observed, due to the gradual change of inertia constant in self-adaptive controllers in comparison to the bang-bang control methods.

Furthermore, as shown Figure 7.16(a), the FESS with the proposed controller improves the frequency nadir from 48.84 Hz in case 1, to 49.27 Hz in case 6. This is achieved by exploiting the damping coefficient immediately after the disturbance occurs, as opposed to case 5, and in combination with quickly reaching a higher inertia constant. As discussed in section 7.1.1, the frequency nadir is determined by both the damping and the inertia, with the damping having a greater influence. After the proposed controller, the controller implemented as case 3 shows the best performance in terms of the frequency nadir, as it is the only method which employs a higher damping coefficient, before reaching the frequency nadir. The highest ROCOF and the worst frequency nadir are observed in case 4, since it does not use the damping coefficient, and the inertia constant is increased slowly.

Moreover, it can be seen in Figure 7.16(a), that using the proposed approach, a better damping of the subsequent frequency oscillations is observed, having almost no overshoot in the frequency, while recovering.

Figure 7.17(a) shows the measured active power of the FESS. To calculate the active power, the FESS voltage and current measurements are collected at its terminal using voltage and current transformers. The instantaneous active power is calculated from the voltage and current measurements in the real-time simulator.

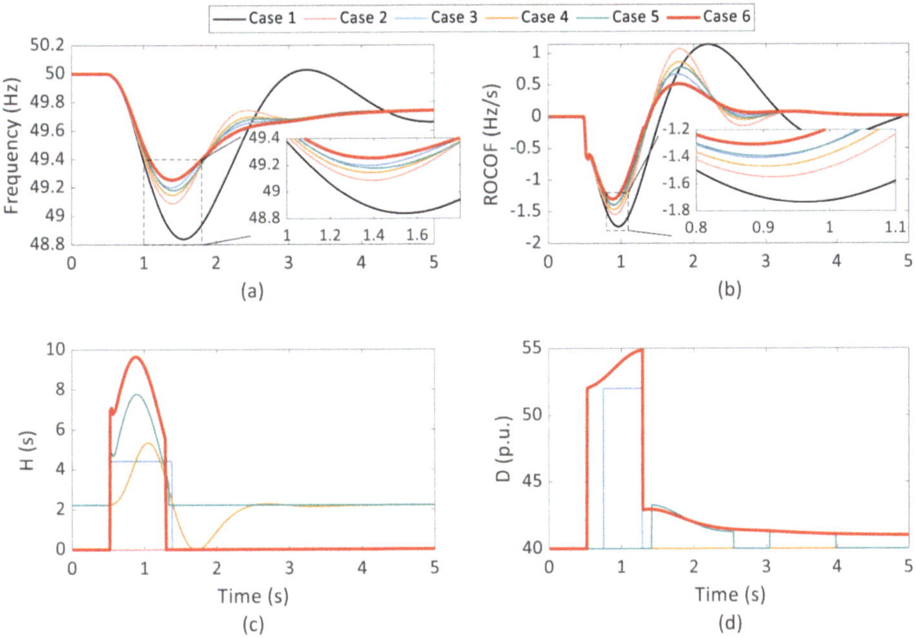

Figure 7.16: PHIL testing results for scenario A: microgrid islanding. (a) Frequency, (b) ROCOF, (c) Inertia constant (H), and (d) Damping coefficient (D).

As seen in Figure 7.17(a), similar to the simulation results, the proposed controller results in a more rapid and prompt response from the FESS to the frequency disturbance, which leads to the improvements observed in the frequency dynamics. The fast response of the FESS is due to the quick change of the inertia constant and damping coefficients using the bang-bang control approach. The active power in case 3 and 6, where the bang-bang approach on both the inertia and the damping are used, reaches a higher peak value. However, as soon as the frequency nadir is reached, the active power is reduced by more than 10 kW and reaches values lower than the power of the FESS in other cases, including the droop controller of case 2. Therefore, the energy drawn from the FESS is not increased significantly. The reduction in the state of charge of the FESS during the five seconds of the islanding scenario is depicted in Figure 7.17(b). In comparison to the conventional droop control, presented as case 2, the proposed inertia controller draws only 0.12 % more energy from the FESS, and only 0.05 %, in comparison to case 3. Therefore, it can be concluded that the improved frequency control using the proposed controller is achieved by only a negligible increase in the energy required from the FESS.

In addition, the PHIL testing results confirm the simulation results in this scenario, regarding the performance of the proposed controller with minor differences in the results. Slightly higher ROCOF values and lower frequency nadirs are observed in the PHIL testing in comparison to the pure simulation studies (see subsection 7.4.5).

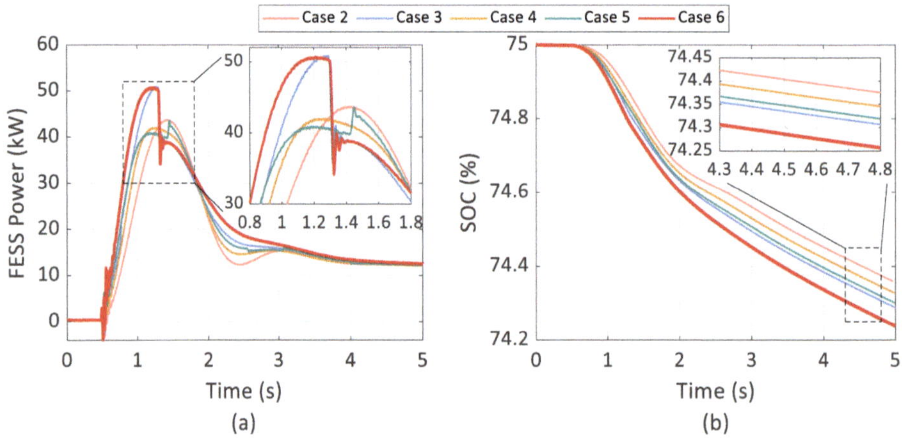

Figure 7.17: PHIL testing results for scenario A: microgrid islanding. (a) Measured FESS power, (b) FESS State of Charge (SOC).

7.4.4 Scenario B: Drop in the PV Generation

In this scenario, a sudden drop in the solar irradiance density is simulated, from 1000 W/m^2 to 150 W/m^2, while the microgrid is disconnected from the mains. The drop in the solar irradiance density, as the main input of the two PV systems, leads to a drastic decrease in the generated power of these systems. With the drop in PV generation, the frequency falls, while the CHP system slowly increases its power to compensate for the power deficit. The FESS attempts to respond as fast as possible to reduce the maximum ROFOC during the disturbance and limit the frequency nadir.

The results of the PHIL testing of the FESS for the six cases in this scenario are shown in Figues 7.18 and 7.19. Similar results in comparison to the simulations are obtained. It can be seen from Figure 7.18(b) that the use of the FESS with an inertia emulation controller, independent of the adaptive inertia and damping method, is able to reduce the ROCOF to below 1 Hz/s. However, the highest reduction in the maximum ROCOF is achieved using the proposed adaptive controller, from 1.16 Hz/s in case 1, to 0.83 Hz/s in case 6. Between the previously reported methods, the controller in case 3 performs better than the others, reducing the ROCOF to 0.9 Hz/s. Unlike scenario A, the controller in case 5 performs worse than the controller in case 3, due to the fact that the severity of the frequency disturbance is smaller in this scenario, and the control parameters in case 5 scale only with the amplitude of the ROCOF and the frequency error.

The frequency variations of the microgrid during this scenario are shown in Figure 7.18(a). It can be observed that the proposed controller improves the frequency nadir from 49.04 Hz in case 1, to 49.31 Hz in case 6, achieving the best results in terms of the frequency nadir. Moreover, a better damping of the frequency oscillations is observed, when the frequency is recovering to its pre-disturbance value.

The measured active power of FESS and its state of charge are shown in Figure 7.19. In comparison to scenario A, the FESS is injecting 12 kW of power, prior to the disturbance, according to damping coefficient D_2. The initial state of charge of the FESS is different in each case, as this scenario has been simulated after an islanding scenario using different controllers in each case. Again, using the proposed adaptive inertia emulation controller, the FESS increases its power more quickly, as seen in Figure 7.19(a), which leads to the improved frequency dynamics of the microgrid. However, the increase in the energy required from the FESS by using the proposed controller is not significant.

It can be concluded that the PHIL testing results of the 60 kW high-speed FESS with the suggested adaptive inertia emulation control confirm the offline simulation results. The proposed controller for the FESS can effectively limit the ROCOF and frequency nadir during an imbalance between generation and demand. Moreover, it outperforms several previously reported methods in literature, while not requiring noticeably more energy from the FESS. The use of the suggested control design requires a higher ramp rate of the active power drawn from the FESS when a disturbance occurs, which is not a problem for a FESS as a storage technology, as it has no effect on its lifetime or capacity.

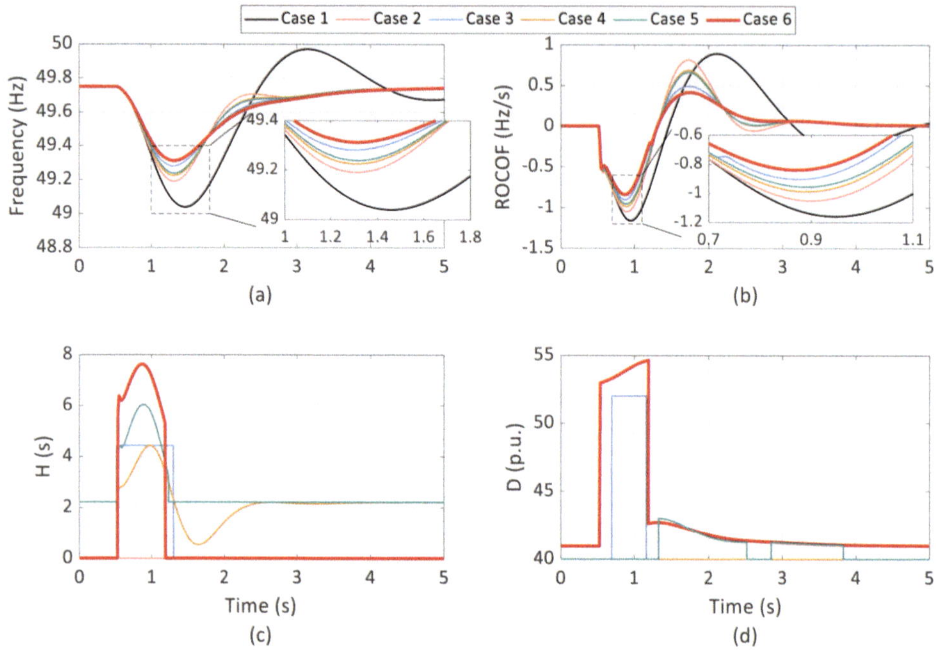

Figure 7.18: PHIL testing results for scenario B: drop in PV generation. (a) Frequency, (b) ROCOF, (c) Inertia constant (H), and (d) Damping coefficient (D).

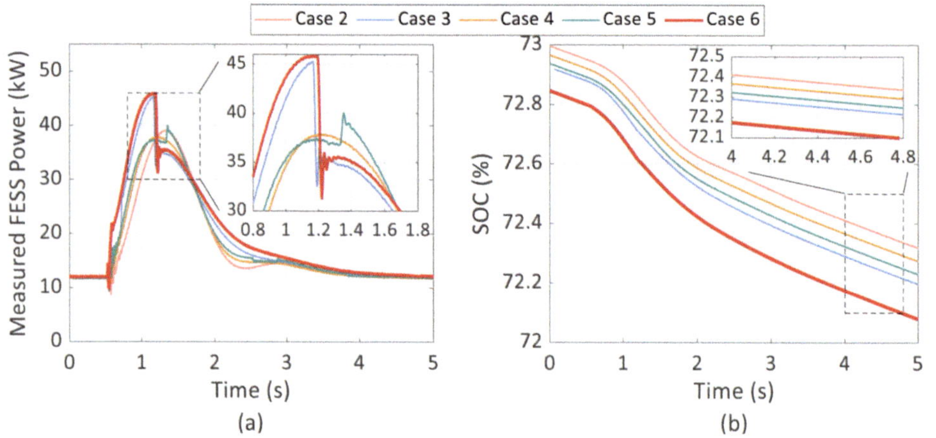

Figure 7.19: PHIL testing results for scenario B: drop in PV generation. (a) Measured FESS power, (b) FESS State of Charge (SOC).

7.4.5 Comparison of the PHIL Testing and the Simulation Results

By comparing the simulation results presented in section 7.3 and the results of the PHIL testing of the real FESS in section 7.4, a good match between the two can be observed. However, slight differences can be seen, which could have been caused by the delays and limitations in the real hardware, which are not considered in the simulated model. Some examples include the nonlinear torque ramp limiter of the permanent magnet synchronous machine and the delays in signal transmission, which exist only in the real hardware, and are difficult to implement in the simulations. These delays and limitations result in slightly higher ROCOF values, lower frequency nadir, and higher power drawn from the FESS. This is a clear advantage of the PHIL testing as a validation tool, where the behavior of a real hardware including all its limitations can be directly reflected on the grid variables, such as the frequency.

In order to see the exact differences between the response of the real hardware and the simulated model of the FESS, the same active power reference, sent to the real FESS during the PHIL testing, is sent simultaneously to the real-time simulation model of the FESS, developed in Chapter 5. The model is evaluated by comparing the measured active power of the real FESS and the simulated one. As an example, the results for the scenario B using the proposed controller (case 6) are shown in Figure 7.20. As seen, there is a good match between the response of the real-time model and the real FESS with a slight difference at the peak power. The difference between the peak power of the model and the real FESS is only 0.93 kW, which corresponds to only 1.8 % of error.

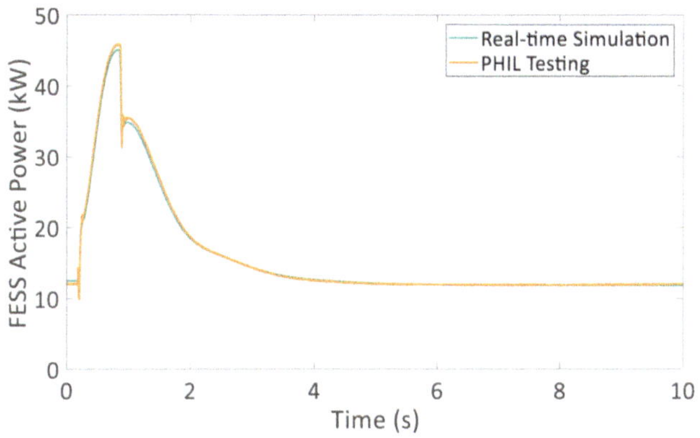

Figure 7.20: A comparison of the real FESS behavior and its simulated model, given the same active power reference from the PHIL testing.

7.5 Summary

Low-inertia power systems can experience a high ROCOF and large frequency deviations during imbalances in generation and demand. In this chapter, a new adaptive inertia emulation controller for a high-speed FESS is introduced, aiming to simultaneously reduce the ROCOF and improve the frequency nadir in low-inertia systems during disturbances. The proposed adaptive method combines the advantages of a bang-bang controller and a self-adaptive one, in order to have a fast and yet proportional response to the frequency disturbances. The proposed control design also has the advantage that it can be easily added as an external control to a commercial FESS with no change required in the existing control systems. Furthermore, the proposed design integrates the state of charge and limitations of the FESS at low rotational speeds into the control architecture.

The performance of the proposed controller has been initially evaluated using numerical simulations in a low voltage microgrid. The controller is also implemented on a real 60 kW high-speed FESS using the concept of rapid control prototyping. The performance of the FESS with the proposed adaptive inertia emulation controller, along with several previously suggested controllers, is investigated using PHIL testing of the FESS and the real-time simulation of the studied microgrid. Both simulation and PHIL results confirm that the proposed adaptive inertia emulation controller can significantly reduce the maximum ROCOF and improve the frequency nadir during large frequency disturbances. Moreover, it outperforms several previously reported interval-based controllers in the investigated scenarios, while not requiring significantly more energy from the FESS.

8 Conclusions and Outlook

8.1 Conclusions

This work investigated the use of a FESS for the application of power smoothing and frequency support in low voltage distribution grids from various perspectives, including its allocation, sizing, modeling and model validation, real-time simulation, and PHIL testing in several scenarios.

Firstly, in order to identify the most suitable location for installing a FESS for power smoothing applications, a new data-driven method for estimating the relative voltage sensitivity to active power changes was introduced. The proposed method uses the mutual information score between the voltage and active power measurements to reflect the relative voltage sensitivity to active power changes at different locations of the grid, without the need of having a grid model. This method has been applied for allocating a FESS in a local grid in southern Germany with more than 1200 installation candidates. It was shown that the proposed method can successfully identify the locations, where the voltage is more sensitive to active power changes, and therefore, can benefit the most from a smoother power profile by using a FESS.

Next, a novel methodology for sizing energy storage systems based on historical measurement data was developed. The main contribution of the proposed method is the use of the motif discovery algorithm for detecting reoccurring consumption patterns in power profiles. The detected patterns are then used as the representative of the whole data set for deriving the energy storage characteristics. The method has been employed for sizing two different types of energy storage system, including a FESS for power smoothing applications. It has been demonstrated that the storage systems with the characteristics derived from only the detected patterns can fulfil their installation purposes such as power smoothing and peak shaving for most of the days throughout the whole the measurement period.

In the following chapters, a dynamic model for a high-speed FESS has been presented and validated with experimental results. The advantage of the presented model is that it incorporates the losses and the auxiliary power requirements of the FESS, which leads to an accurate estimation of the state of charge of the FESS during idle, charging, and discharging operational modes of the FESS, in all the investigated scenarios. The maximum difference between the state of charge of the developed model and the real FESS, among all investigated scenarios, has been approximately 0.8 % only.

For verifying the performance of the FESS, a laboratory setup for PHIL testing of a 60 kW high-speed FESS has been built and successfully tested. The setup allows testing the FESS in various scenarios and grid configurations for different applications, and also controlling and monitoring the FESS during the experiments. In order to evaluate the response of the FESS for the application frequency support using this setup, several frequency deviation scenarios including a grid code compliance verification test and the major frequency disturbance of August 9, 2019, in the power system of the UK have been simulated in real-time. It has been demonstrated that the FESS can respond very quickly to frequency deviations, and reaches the required power indicated by the grid code in just under 60 ms.

In order to demonstrate the advantages of the rapid response of the FESS for low-inertia power systems, the use of a FESS for providing emulated inertia in a low voltage microgrid has been investigated. A novel adaptive controller for inertia emulation using a FESS has been introduced, and its performance has been verified using simulations and experiments in two different scenarios in a low-inertia microgrid. The proposed controller combines the advantages of the fast reaction time of bang-bang controllers with the flexibility of self-adaptive ones in order to improve the inertia support provided by the FESS. It also considers the state of charge of the FESS and its power limitations at low rotational speeds into the control structure. Simulation results indicate that during the islanding of the studied microgrid, the use of the FESS with the proposed adaptive inertia emulation controller reduces the maximum ROCOF by 28 %, and the maximum frequency deviation by 44 %, in comparison to the case without the FESS. Also, the proposed controller has demonstrated a better performance in terms of reducing the maximum ROCOF and improving the frequency nadir in comparison to several previously reported adaptive controllers. The suggested controller has also been implemented on the 60 kW high-speed FESS using the concept of rapid control prototyping, and the performance of the FESS equipped with the proposed inertia emulation controller has been verified by means PHIL testing. A good match between the simulation and PHIL testing results has been observed, indicating the effectiveness of the proposed controller and the accuracy of the FESS model. However, slightly higher ROCOF values are observed during the PHIL testing, due to the minor delays present in the real FESS, which demonstrates the advantage of PHIL testing, where real component behaviours are reflected in the grid variables. It is important to note that the results presented in this work are the very first experimental validation of inertia emulation using a FESS.

In summary, this thesis provided a comprehensive study on the grid integration of a high-speed FESS, showing its advantages, such as a quick response time to power imbalances, and how new tools such as PHIL testing and data-driven methods can be used effectively for conducting grid integration studies.

8.2 Outlook

There are several areas where the presented work in this thesis can be extended in the future. With the increasing number of measurement devices in power systems (e.g., smart meters, phasor measurement units), the data-driven approaches introduced for the allocation and sizing of FESS have a great potential to be applied to other use cases and applications. When measurement devices are available at different locations of the grid, the proposed allocation procedure based on the mutual information score can be fully automated, and applied for the allocation of other distributed energy resources. Furthermore, a new method has been recently introduced to estimate the mutual information score in real-time using a sliding window, by updating only a fixed number of nearest-neighbor relationships [175]. Therefore, the use of mutual information, as a measure of linear and nonlinear dependency between two variables, can be investigated for several other applications, such as real-time topology change detection (see [306]), improving load modelling, and adaptive controllers, such as the controller proposed in [169].

The use of the motif discovery algorithm for extracting reoccurring patterns in active power measurements can drastically reduce the input data size, and therefore, the computation time required for solving complex nonlinear optimization problems, which are commonly used for component sizing and system planning. This can be significantly beneficial, in particular when the optimization problem requires data over longer horizons, such as several years. As a future work, it would be interesting to quantify the impact of using motif discovery on the computation time of such optimization frameworks and on their outcome, and compare it to commonly used methods, such as clustering techniques.

The validated model of the FESS, which can accurately reflect its self-discharge rate, can be used by other researchers for simulating and investigating the advantages of a FESS in other applications, or designing and validation new controller. Since the model can be easily simulated in real-time using commercial real-time simulators, it can also be used for Controller Hardware-in-the-Loop (CHIL) testing of new controllers and energy management systems. Moreover, this model together with a power amplifier, that is capable to operate as a current source, can be used to build a FESS emulator, which is a cheaper and safer alternative of having the real FESS. The modeling can be improved in the future by disaggregating different losses and the power of each auxiliary component. This can help to build a more efficient FESS or operating it with less power requirements. Also, the small delays found within the real FESS can be added to the model to improve its accuracy in the millisecond range.

But more importantly, the laboratory setup established within this work for the PHIL testing of the FESS, can be used for evaluating the performance of a real FESS for other applications, including power smoothing and ramp rate control of renewables, voltage regulation, and supporting pulsed power loads in various low voltage grids and microgrids. In addition, the combination of the FESS with other storage technologies with a higher energy density, such as lithium-ion batteries, or with distributed energy resources which operate more efficiently under a constant operating point (e.g., fuel cells, micro gas turbines), can be investigated with the real FESS using this setup. As shown, this setup also enables to test new control strategies or energy management systems on the real FESS for each application, using the concept of rapid control prototyping. By combining rapid control prototyping and PHIL testing, this setup allows to verify the performance of these new controllers under the practical limitations and constraints of a real FESS, which are very often neglected in numerical simulation studies found in literature. These include auxiliary power requirements and losses of the FESS, its power limitations at low rotational speeds, delays and other system nonlinearities.

While the FESS has proven to be a viable solution for short-term high-power applications, further improvements in its efficiency (for instance, by using more efficient power converters, electrical machines, and bearings), reduction in its auxiliary power requirements, and a more competitive capital costs can potentially lead to an increased presence of this technology in power systems, in particular in low-inertia systems.

Bibliography

[1] International Energy Agency (IEA), "Renewables 2020: Analysis and Forecast to 2025," Paris, France, 2020.

[2] International Renewable Energy Agency (IRENA), "Electricity storage and renewables: Costs and markets to 2030," Abu Dhabi, UAE, 2017.

[3] Federal Ministry for Economic Affairs and Energy, "BMWi - Federal Ministry for Economic Affairs and Energy - Renewable Energy," 2020. [Online]. Available: https://www.bmwi.de/Redaktion/EN/Dossier/renewable-energy.html. [Accessed: 17-Nov-2020].

[4] C. Zhang, Y. L. Wei, P. F. Cao, and M. C. Lin, "Energy storage system: Current studies on batteries and power condition system," *Renewable and Sustainable Energy Reviews*, vol. 82, no. July, pp. 3091–3106, 2018.

[5] ENTSO-E, "Frequency Stability Evaluation Criteria for the Synchronous Zone of Continental Europe," Brussels, Belgium, 2016.

[6] ENTSO-E, "High Penetration of Power Electronic Interfaced Power Sources and the Potential Contribution of Grid Forming Converters," Brussels, Belgium, 2019.

[7] L. Meng, J. Zafar, S. K. Khadem, A. Collinson, K. C. Murchie, F. Coffele, and G. M. Burt, "Fast Frequency Response from Energy Storage Systems - A Review of Grid Standards, Projects and Technical Issues," *IEEE Transactions on Smart Grid*, vol. 11, no. 2, pp. 1566–1581, 2020.

[8] International Energy Agency (IEA), "Technology Roadmap: Energy Storage," Paris, France, 2014.

[9] Electric Power Research Institute (EPRI), "Electric Energy Storage Technology Options: A White Paper Primer on Applications, Costs and Benefits," Palo Alto, California, 2010.

[10] D. M. Greenwood, K. Y. Lim, C. Patsios, P. F. Lyons, Y. S. Lim, and P. C. Taylor, "Frequency response services designed for energy storage," *Applied Energy*, vol. 203, pp. 115–127, 2017.

[11] G. Delille, B. François, and G. Malarange, "Dynamic frequency control support by energy storage to reduce the impact of wind and solar generation on isolated power system's inertia," *IEEE Transactions on Sustainable Energy*, vol. 3, no. 4, pp. 931–939, 2012.

[12] U. Akram, M. Nadarajah, R. Shah, and F. Milano, "A review on rapid responsive energy storage technologies for frequency regulation in modern power systems," *Renewable and Sustainable Energy Reviews*, vol. 120, p. 109626, 2020.

[13] B. Zakeri and S. Syri, "Electrical energy storage systems: A comparative life cycle cost analysis," *Renewable and Sustainable Energy Reviews*, vol. 42, pp. 569–596, 2015.

[14] M. Farhadi and O. Mohammed, "Energy Storage Technologies for High-Power Applications," *IEEE Transactions on Industry Applications*, vol. 52, no. 3, pp. 1953–1962, 2016.

[15] D. I. Stroe, V. Knap, M. Swierczynski, A. I. Stroe, and R. Teodorescu, "Operation of a grid-connected lithium-ion battery energy storage system for primary frequency regulation: A battery lifetime perspective," *IEEE Transactions on Industry Applications*, vol. 53, no. 1, pp. 430–438, 2017.

[16] K. R. Pullen, "The Status and Future of Flywheel Energy Storage," *Joule*, vol. 3, no. 6, pp. 1394–1399, 2019.

[17] P. Keil, "Aging of Lithium-Ion Batteries in Electric Vehicles," Technische Universität München, 2017.

[18] P. C. Kotsampopoulos, F. Lehfuss, G. F. Lauss, B. Bletterie, and N. D. Hatziargyriou, "The limitations of digital simulation and the advantages of PHIL testing in studying distributed generation provision of ancillary services," *IEEE Transactions on Industrial Electronics*, vol. 62, no. 9, pp. 5502–5515, 2015.

[19] D. M. Greenwood, K. Y. Lim, C. Patsios, P. F. Lyons, Y. S. Lim, and P. C. Taylor, "Frequency response services designed for energy storage," *Applied Energy*, vol. 203, pp. 115–127, 2017.

[20] A. ; Pratt, J. ; Wang, K. ; Prabakar, M. M. ; Baggu, M. ; Symko-Davies, N. ; Ainsworth, and T. Bialek, "Evaluating the Impacts of Battery Energy Storage System Functionalities on Distribution Systems Using Power Hardware-in-the-Loop Simulation," in *2018 IEEE Power and Energy Society General Meeting*, Portland, Oregon, 2018.

[21] Y. J. Kim and J. Wang, "Power Hardware-in-the-Loop Simulation Study on Frequency Regulation Through Direct Load Control of Thermal and Electrical Energy Storage Resources," *IEEE Transactions on Smart Grid*, vol. 9, no. 4, pp. 2786–2796, 2018.

[22] Y. Wang, G. Delille, X. Guillaud, F. Colas, and B. François, "Real-time simulation: The missing link in the design process of advanced grid equipment," in *2010 IEEE PES General Meeting*, Minneapolis, MN, USA, 2010.

[23] S. Kochanneck, I. Mauser, B. Bohnet, S. Hubschneider, H. Schmeck, M. Braun,

and T. Leibfried, "Establishing a hardware-in-the-loop research environment with a hybrid energy storage system," in *IEEE PES Innovative Smart Grid Technologies Conference Europe*, Melbourne, 2016, pp. 497–503.

[24] A. Downie, A. Avras, K. Jennett, and F. Coffele, "Power Hardware in the Loop Platform for Flywheel Energy Storage System Testing for Electric Ship Power System Applications," in *2nd International Conference on Modelling and Optimisation of Ship Energy Systems*, Glasgow, Scotland, 2019.

[25] K. I. Jennett, A. S. H. Downie, A. Avras, F. Coffele, A. Tate, and S. Lewinton, "Performance Optimisation of a Flywheel Energy Storage System using the PNDC Power Hardware in the Loop Platform," in *Proceedings of the Engine As A Weapon International Symposium (EAAW)*, London, UK, 2019.

[26] T. M. Cover and J. A. Thomas, *Elements of Information Theory*, Second. Hoboken, New Jersey: John Wiley & Sons, Inc, 2006.

[27] B. Chiu, E. Keogh, and S. Lonardi, "Probabilistic discovery of time series motifs," in *Proceedings of the ninth ACM SIGKDD international conference on Knowledge discovery and data mining - KDD '03*, New York, NY, USA, 2003, pp. 493–498.

[28] R. Sebastián and R. Peña Alzola, "Flywheel energy storage systems: Review and simulation for an isolated wind power system," *Renewable and Sustainable Energy Reviews*, vol. 16, no. 9, pp. 6803–6813, 2012.

[29] G. Holler, "Flywheel systems in automotive applications: Design criteria, challenges, and effects on mobile energy storage systems," in *Eco-Mobility Conference*, Vienna, Austria, 2012.

[30] X. Tan, Q. Li, and H. Wang, "Advances and trends of energy storage technology in Microgrid," *International Journal of Electrical Power and Energy Systems*, vol. 44, no. 1, pp. 179–191, 2013.

[31] R. H. Richman, "Flywheel Energy Storage," Palo Alto, California, 1997.

[32] International Energy Agency (IEA), "Energy Storage – Analysis - IEA," 2020. [Online]. Available: https://www.iea.org/reports/energy-storage. [Accessed: 12-Nov-2020].

[33] Z. Zhou, M. Benbouzid, J. Frédéric Charpentier, F. Scuiller, and T. Tang, "A review of energy storage technologies for marine current energy systems," *Renewable and Sustainable Energy Reviews*, vol. 18. pp. 390–400, 2013.

[34] M. E. Amiryar, K. R. Pullen, and D. Nankoo, "Development of a high-fidelity model for an electrically driven energy storage flywheel suitable for small scale residential applications," *Applied Sciences*, vol. 8, no. 3, pp. 1–29, 2018.

[35] N. Ghaviha, J. Campillo, M. Bohlin, and E. Dahlquist, "Review of Application of Energy Storage Devices in Railway Transportation," *Energy Procedia*, vol. 105, pp. 4561–4568, 2017.

[36] B. Fan, C. Wang, Q. Yang, W. Liu, and G. Wang, "Performance Guaranteed Control of Flywheel Energy Storage System for Pulsed Power Load Accommodation," *IEEE Transactions on Power Systems*, vol. 33, no. 4, pp. 3994–4004, 2018.

[37] B. H. Kenny and P. E. Kascak, "Sensorless Control of Permanent Magnet Machine for NASA Flywheel Technology Development," in *37th Intersociety Energy Conversion Engineering Conference*, Washington, DC, 2002.

[38] M. L. Lazarewicz and T. M. Ryan, "Integration of flywheel-based energy storage for frequency regulation in deregulated markets," in *IEEE PES General Meeting 2010*, Providence, RI, USA, 2010.

[39] A. Colthorpe, "Flywheel-lithium battery hybrid energy storage system joining Dutch grid services markets | Energy Storage News," 2020. [Online]. Available: https://www.energy-storage.news/news/flywheel-lithium-battery-hybrid-energy-storage-system-joining-dutch-grid-se. [Accessed: 13-Nov-2020].

[40] H. Silva-Saravia, H. Pulgar-Painemal, and J. M. Mauricio, "Flywheel Energy Storage Model, Control and Location for Improving Stability: The Chilean Case," *IEEE Transactions on Power Systems*, vol. 32, no. 4, pp. 3111–3119, 2017.

[41] N. Hamsic, A. Schmelter, A. Mohd, E. Ortjohann, E. Schultze, A. Tuckey, and J. Zimmermann, "Stabilising the Grid Voltage and Frequency in Isolated Power Systems Using a Flywheel Energy Storage System," in *The Great Wall World Renewable Energy Forum*, Beijing, China, 2006.

[42] M. S. Mahdavi, G. B. Gharehpetian, and H. A. Moghaddam, "Enhanced Frequency Control Method for Microgrid-Connected Flywheel Energy Storage System," *IEEE Systems Journal (Early Access)*, pp. 1–11, 2020.

[43] A. A. Khodadoost Arani, G. B. Gharehpetian, and M. Abedi, "A Novel Control Method Based on Droop for Cooperation of Flywheel and Battery Energy Storage Systems in Islanded Microgrids," *IEEE Systems Journal*, vol. 14, no. 1, pp. 1080–1087, 2020.

[44] F. Sanchez, J. Cayenne, F. Gonzalez-Longatt, and J. L. Rueda, "Controller to enable the enhanced frequency response services from a multi-electrical energy storage system," *IET Generation, Transmission and Distribution*, vol. 13, no. 2, pp. 258–265, 2019.

[45] J. Yu, J. Fang, and Y. Tang, "Inertia Emulation by Flywheel Energy Storage System for Improved Frequency Regulation," in *2018 IEEE 4th Southern Power Electronics Conference, SPEC 2018*, Singapore, Singapore, 2018.

[46] R. Peña-Alzola, D. Campos-Gaona, and M. Ordonez, "Control of flywheel energy storage systems as virtual synchronous machines for microgrids," in *2015 IEEE 16th Workshop on Control and Modeling for Power Electronics, COMPEL 2015*, Vancouver, BC, Canada, 2015.

[47] A. Maitra, L. Rogers, and R. Handa, "Program on Technology Innovation: Microgrid Implementations: Literature Review," Palo Alto, California, 2016.

[48] A. S. Mir and N. Senroy, "Intelligently Controlled Flywheel Storage for Enhanced Dynamic Performance," *IEEE Transactions on Sustainable Energy*, vol. 10, no. 4, pp. 2163–2173, 2019.

[49] S. Ghosh and S. Kamalasadan, "An Energy Function-Based Optimal Control Strategy for Output Stabilization of Integrated DFIG-Flywheel Energy Storage System," *IEEE Transactions on Smart Grid*, vol. 8, no. 4, pp. 1922–1931, 2017.

[50] G. Cimuca, S. Breban, M. M. Radulescu, C. Saudemont, and B. Robyns, "Design and control strategies of an induction-machine-based flywheel energy storage system associated to a variable-speed wind generator," *IEEE Transactions on Energy Conversion*, vol. 25, no. 2, pp. 526–534, 2010.

[51] L. Wang, J.-Y. Yu, and Y.-T. Chen, "Dynamic stability improvement of an integrated offshore wind and marine-current farm using a flywheel energy-storage system," *IET Renewable Power Generation*, vol. 5, no. 2, pp. 387–396, 2011.

[52] F. Islam, A. Al-Durra, and S. M. Muyeen, "Smoothing of wind farm output by prediction and supervisory-control-unit- based FESS," *IEEE Transactions on Sustainable Energy*, vol. 4, no. 4, pp. 925–933, 2013.

[53] R. Peña-Alzola, R. Sebastián, J. Quesada, and A. Colmenar, "Review of flywheel based energy storage systems," in *International Conference on Power Engineering, Energy and Electrical Drives*, Torremolinos (Málaga), Spain, 2011.

[54] "ABB Helping To Power Australia's Outback With Solar," 2015. [Online]. Available: https://www.energymatters.com.au/renewable-news/abb-australia-solar-em5033/. [Accessed: 11-Nov-2020].

[55] L. Barelli, G. Bidini, F. Bonucci, L. Castellini, A. Fratini, F. Gallorini, and A. Zuccari, "Flywheel hybridization to improve battery life in energy storage systems coupled to RES plants," *Energy*, vol. 173, pp. 937–950, 2019.

[56] P. Mouratidis, B. Schubler, and S. Rinderknecht, "Hybrid energy storage system consisting of a flywheel and a lithium-ion battery for the provision of primary control reserve," in *8th International Conference on Renewable Energy Research and Applications, ICRERA 2019*, Brasov, Romania, 2019, no. 03, pp. 94–99.

[57] J. Hou, J. Sun, and H. Hofmann, "Control development and performance evaluation for battery/flywheel hybrid energy storage solutions to mitigate load

fluctuations in all-electric ship propulsion systems," *Applied Energy*, vol. 212, pp. 919–930, 2018.

[58] P. Fairley, "Flywheels Get their Spin Back With Beacon Power's Rebound," *IEEE Spectrum*, 2014. [Online]. Available: https://spectrum.ieee.org/energy/fossil-fuels/flywheels-get-their-spin-back-with-beacon-powers-rebound. [Accessed: 13-Jul-2020].

[59] R. S. Weissbach, G. G. Karady, and R. G. Farmer, "Dynamic voltage compensation on distribution feeders using flywheel energy storage," *IEEE Transactions on Power Delivery*, vol. 14, no. 2, pp. 465–471, 1999.

[60] B. Wang and G. Venkataramanan, "Dynamic voltage restorer utilizing a matrix converter and flywheel energy storage," *IEEE Transactions on Industry Applications*, vol. 45, no. 1, pp. 222–231, 2009.

[61] P. Gambôa, J. F. Silva, S. F. Pinto, and E. Margato, "Predictive optimal matrix converter control for a dynamic voltage restorer with flywheel energy storage," in *2009 35th Annual Conference of IEEE Industrial Electronics*, Porto, Portugal, 2009, pp. 759–764.

[62] A. Al-Diab and C. Sourkounis, "Unbalanced voltage drops compensations using flywheel energy storage system," in *Proceeding of the International Conference on Electrical Power Quality and Utilisation, EPQU*, Lisbon, Portugal, 2011, pp. 618–623.

[63] M. I. Daoud, A. Massoud, S. Ahmed, A. S. Abdel-Khalik, and A. Elserougi, "Ride-through capability enhancement of VSC-HVDC based wind farms using low speed flywheel energy storage system," in *Conference Proceedings - IEEE Applied Power Electronics Conference and Exposition - APEC*, Fort Worth, TX, USA, 2014.

[64] M. H. Wang and H. C. Chen, "Transient stability control of multimachine power systems using flywheel energy injection," *IEE Proceedings - Generation, Transmission and Distribution*, vol. 152, no. 5, pp. 589–596, 2005.

[65] T. Dragičević, S. Sučić, J. C. Vasquez, and J. M. Guerrero, "Flywheel-based distributed bus signalling strategy for the public fast charging station," *IEEE Transactions on Smart Grid*, vol. 5, no. 6, pp. 2825–2835, 2014.

[66] B. Bolund, H. Bernhoff, and M. Leijon, "Flywheel energy and power storage systems," *Renewable and Sustainable Energy Reviews*, vol. 11, no. 2, pp. 235–258, 2007.

[67] S. M. Mousavi G, F. Faraji, A. Majazi, and K. Al-Haddad, "A comprehensive review of Flywheel Energy Storage System technology," *Renewable and Sustainable Energy Reviews*, vol. 67, pp. 477–490, 2017.

[68] "STORNETIC - The Energy Storage Company." [Online]. Available: https://stornetic.com/. [Accessed: 12-Nov-2020].

[69] M. E. Amiryar and K. R. Pullen, "A review of flywheel energy storage system technologies and their applications," *Applied Sciences*, vol. 7, no. 3, p. 286, 2017.

[70] X. Li, B. Anvari, A. Palazzolo, Z. Wang, and H. Toliyat, "A Utility-Scale Flywheel Energy Storage System with a Shaftless, Hubless, High-Strength Steel Rotor," *IEEE Transactions on Industrial Electronics*, vol. 65, no. 8, pp. 6667–6675, 2018.

[71] "The Next Frontier in Energy Storage | Amber Kinetics." [Online]. Available: https://www.amberkinetics.com/product/. [Accessed: 20-Nov-2020].

[72] L. Jiang, W. Zhang, G. J. Ma, and C. W. Wu, "Shape optimization of energy storage flywheel rotor," *Structural and Multidisciplinary Optimization*, vol. 55, no. 2, pp. 739–750, 2017.

[73] P. C. Krause, O. Wasynczuk, S. D. Sudhoff, and S. Pekarek, *Analysis of Electric Machinery and Drive Systems*, Third Edit. Hoboken, New Jersey: Wiley-IEEE Press, 2010.

[74] L. Uršič and M. Nemec, "Permanent magnet synchronous machine demagnetisation prevention and torque estimation control considering rotor temperature," *IET Power Electronics*, vol. 12, no. 9, pp. 2161–2169, 2019.

[75] C. Y. Ho, J. C. Wang, K. W. Hu, and C. M. Liaw, "Development and Operation Control of a Switched-Reluctance Motor Driven Flywheel," *IEEE Transactions on Power Electronics*, vol. 34, no. 1, pp. 526–537, 2019.

[76] S. J. Amodeo, H. G. Chiacchiarini, J. A. Solsona, and C. A. Busada, "High-performance sensorless nonlinear power control of a flywheel energy storage system," *Energy Conversion and Management*, vol. 50, no. 7, pp. 1722–1729, 2009.

[77] T. D. Nguyen, G. Foo, K. J. Tseng, and D. M. Vilathgamuwa, "Modeling and sensorless direct torque and flux control of a dual-airgap axial flux permanent-magnet machine with field-weakening operation," *IEEE/ASME Transactions on Mechatronics*, vol. 19, no. 2, pp. 412–422, 2014.

[78] S. Koohi-Kamali, V. V. Tyagi, N. A. Rahim, N. L. Panwar, and H. Mokhlis, "Emergence of energy storage technologies as the solution for reliable operation of smart power systems: A review," *Renewable and Sustainable Energy Reviews*, vol. 25, pp. 135–165, 2013.

[79] A. A. K. Arani, H. Karami, G. B. Gharehpetian, and M. S. A. Hejazi, "Review of Flywheel Energy Storage Systems structures and applications in power systems and microgrids," *Renewable and Sustainable Energy Reviews*, vol. 69, pp. 9–18,

2017.

[80] M. Mansour, M. N. Mansouri, and M. F. Mimouni, "Performance evaluation of a flywheel energy-storage system associated to a variable-speed wind generator," *IEEE Transactions on Smart Grid*, vol. 53, no. 4, pp. 1074–1085, 2006.

[81] B. Y. A. Ausko and J. Dedad, "Stored energy: Short-term and long-term energy storage methods for standby electric power system," *IEEE Industry Applications Magazine*, vol. 13, no. 4, pp. 66–72, 2007.

[82] N. B. B. Ahamad, C. L. Su, X. Zhaoxia, J. C. Vasquez, J. M. Guerrero, and C. H. Liao, "Energy Harvesting from Harbor Cranes with Flywheel Energy Storage Systems," *IEEE Transactions on Industry Applications*, vol. 55, no. 4, pp. 3354–3364, 2019.

[83] X. Chang, Y. Li, W. Zhang, N. Wang, and W. Xue, "Active disturbance rejection control for a flywheel energy storage system," *IEEE Transactions on Industrial Electronics*, vol. 62, no. 2, pp. 991–1001, 2015.

[84] A. De Bernardinis, A. Kolli, J. P. Ousten, and R. Lallemand, "High efficiency Silicon carbide DC-AC inverter for EV-charging flywheel system," in *2017 IEEE Transportation and Electrification Conference and Expo, ITEC 2017*, Chicago, IL, USA, 2017, pp. 421–424.

[85] W. Santiago, "Inverter output filter effect on PWM motor drives of a flywheel energy storage system," in *2nd International Energy Conversion Engineering Conference*, Providence, Rhode Island, USA, 2004, pp. 914–932.

[86] G. O. Suvire, M. G. Molina, and P. E. Mercado, "Improving the integration of wind power generation into AC microgrids using flywheel energy storage," *IEEE Transactions on Smart Grid*, vol. 3, no. 4, pp. 1945–1954, 2012.

[87] S. J. Amodeo, H. G. Chiacchiarini, and A. R. Oliva, "High-performance control of a DC-DC Z-source converter used for an excitation field driver," *IEEE Transactions on Power Electronics*, vol. 27, no. 6, pp. 2947–2957, 2012.

[88] S. R. Gurumurthy, V. Agarwal, and A. Sharma, "High-Efficiency Bidirectional Converter for Flywheel Energy Storage Application," *IEEE Transactions on Industrial Electronics*, vol. 63, no. 9, pp. 5477–5487, 2016.

[89] M. Murayama, S. Kato, H. Tsutsui, S. Tsuji-Iio, and R. Shimada, "Combination of flywheel energy storage system and boosting modular multilevel cascade converter," *IEEE Transactions on Applied Superconductivity*, vol. 28, no. 3, pp. 2018–2021, 2018.

[90] M. Strasik, P. E. Johnson, A. C. Day, J. Mittleider, M. D. Higgins, J. Edwards, J. R. Schindler, K. E. McCrary, C. R. McIver, D. Carlson, J. F. Gonder, and J. R. Hull, "Design, fabrication, and test of a 5-kWh/100-kW flywheel energy storage

utilizing a high-temperature superconducting bearing," *IEEE Transactions on Applied Superconductivity*, vol. 17, no. 2, pp. 2133–2137, 2007.

[91] T. Matsushita, *Flux pinning in superconductors*, Second Edi. Springer Science, 2007.

[92] M. Komori and N. Akinaga, "A prototype of flywheel energy storage system suppressed by hybrid magnetic bearings with H∞ controller," *IEEE Transactions on Applied Superconductivity*, vol. 11, no. 1, pp. 1733–1736, 2001.

[93] T. M. Mulcahy, J. R. Hull, K. L. Uherka, R. G. Abboud, and J. J. Juna, "Test results of 2-kWh flywheel using passive PM and HTS bearings," *IEEE Transactions on Applied Superconductivity*, vol. 11, no. 1, pp. 1729–1732, 2001.

[94] F. N. Werfel, U. Floegel-Delor, T. Riedel, R. Rothfeld, D. Wippich, B. Goebel, G. Reiner, and N. Wehlau, "Towards high-capacity HTS flywheel systems," *IEEE Transactions on Applied Superconductivity*, vol. 20, no. 4, pp. 2272–2275, 2010.

[95] A. D. Sapowith and W. E. Handy, "Composite-flywheel burst-containment study," Livermore, CA, 1982.

[96] J. Belanger, P. Venne, and J.-N. Paquin, "The What, Where and Why of Real-time Simulation," *RT Planet*, 2010.

[97] M. D. Omar Faruque, T. Strasser, G. Lauss, V. Jalili-Marandi, P. Forsyth, C. Dufour, V. Dinavahi, A. Monti, P. Kotsampopoulos, J. A. Martinez, K. Strunz, M. Saeedifard, Xiaoyu Wang, D. Shearer, and M. Paolone, "Real-Time Simulation Technologies for Power Systems Design, Testing, and Analysis," *IEEE Power and Energy Technology Systems Journal*, vol. 2, no. 2, pp. 63–73, 2015.

[98] G. F. Lauss, M. O. Faruque, K. Schoder, C. Dufour, A. Viehweider, and J. Langston, "Characteristics and design of power hardware-in-The-loop simulations for electrical power systems," *IEEE Transactions on Industrial Electronics*, vol. 63, no. 1, pp. 406–417, 2016.

[99] A. Rasheed, O. San, and T. Kvamsdal, "Digital twin: Values, challenges and enablers from a modeling perspective," *IEEE Access*, vol. 8, pp. 21980–22012, 2020.

[100] X. Guillaud, M. O. Faruque, A. Teninge, A. H. Hariri, L. Vanfretti, M. Paolone, V. Dinavahi, P. Mitra, G. Lauss, C. Dufour, P. Forsyth, A. K. Srivastava, K. Strunz, T. Strasser, and A. Davoudi, "Applications of Real-Time Simulation Technologies in Power and Energy Systems," *IEEE Power and Energy Technology Systems Journal*, vol. 2, no. 3, pp. 103–115, 2015.

[101] A. Yamane, S. Abourida, Y. Bouzid, and F. Tempez, "Real-Time Simulation of Distributed Energy Systems and Microgrids," in *IFAC-PapersOnLine*, 2016, vol. 49, no. 27, pp. 183–187.

[102] P. Forsyth, O. Nzimako, C. Peters, and M. Moustafa, "Challenges of modeling electrical distribution networks in real-time," in *Proceedings - 2015 International Symposium on Smart Electric Distribution Systems and Technologies (EDST)*, Vienna, Austria, 2015, pp. 556–559.

[103] A. Teninge, Y. Besanger, F. Colas, H. Fakham, and X. Guillaud, "Real-time simulation of a medium scale distribution network: Decoupling method for multi-CPU computation," in *2012 IEEE Workshop on Complexity in Engineering, COMPENG 2012*, Aachen, Germany, 2012, pp. 77–82.

[104] H. Hooshyar, L. Vanfretti, and C. Dufour, "Delay-free parallelization for real-time simulation of a large active distribution grid model," in *Industrial Electronics Conference (IECON)*, Florence, Italy, 2016, pp. 6278–6284.

[105] C. Dufour and J. Belanger, "On the use of real-time simulation technology in smart grid research and development," *IEEE Transactions on Industry Applications*, vol. 50, no. 6, pp. 3963–3970, Nov. 2014.

[106] M. Dyck and O. Nzimako, "Real-time simulation of large distribution networks with distributed energy resources," *CIRED - Open Access Proceedings Journal*, vol. 2017, no. 1, pp. 1402–140, 2017.

[107] H. Saad, S. Dennetière, J. Mahseredjian, T. Ould-Bachir, and J. P. David, *Simulation of Transients for VSC-HVDC Transmission Systems Based on Modular Multilevel Converters*. Chichester: Wiley-IEEE Press, 2014.

[108] J. Mahseredjian, V. Dinavahi, and J. A. Martinez, "Simulation tools for electromagnetic transients in power systems: Overview and challenges," *IEEE Transactions on Power Delivery*, vol. 24, no. 3, pp. 1657–1669, 2009.

[109] C. Dufour, J. Mahseredjian, and J. Bélanger, "A combined state-space nodal method for the simulation of power system transients," *IEEE Transactions on Power Delivery*, vol. 26, no. 2, pp. 928–935, 2011.

[110] G. De Carne, M. Langwasser, M. Ndreko, R. Bachmann, R. W. De Doncker, R. Dimitrovski, B. J. Mortimer, A. Neufeld, F. Rojas, and M. Liserre, "Which Deepness Class Is Suited for Modeling Power Electronics?: A Guide for Choosing the Right Model for Grid-Integration Studies," *IEEE Industrial Electronics Magazine*, vol. 13, no. 2, pp. 41–55, 2019.

[111] W. Li, G. Joós, and J. Bélanger, "Real-time simulation of a wind turbine generator coupled with a battery supercapacitor energy storage system," *IEEE Transactions on Industrial Electronics*, vol. 57, no. 4, pp. 1137–1145, 2010.

[112] F. Vasca and L. Iannelli, *Dynamics and Control of Switched Electronic Systems*. London: Springer, 2013.

[113] W. Ren, M. Sloderbeck, M. Steurer, V. Dinavahi, T. Noda, S. Filizadeh, A. R.

Chevrefils, M. Matar, R. Iravani, C. Dufour, J. Belanger, M. O. Faruque, K. Strunz, and J. A. Martinez, "Interfacing issues in real-time digital simulators," *IEEE Transactions on Power Delivery*, vol. 26, no. 2, pp. 1221–1230, 2011.

[114] M. Sloderbeck, R. Meka, M. O. Faruque, J. Langston, and M. Steurer, "Co-simulation of an FPGA-based electromagnetic transient model and a small time-step model in the RTDS real-time digital simulator," in *2012 Grand Challenges in Modeling and Simulation (GCMS 2012)*, 2012, pp. 100–104.

[115] L. A. Gregoire, H. F. Blanchette, J. Belanger, and K. Al-Haddad, "Real-time simulation-based multisolver decoupling technique for complex power-electronics circuits," *IEEE Transactions on Power Delivery*, vol. 31, no. 5, pp. 2313–2321, Oct. 2016.

[116] T. Ould-Bachir, H. F. Blanchette, and K. Al-Haddad, "A network tearing technique for FPGA-based real-time simulation of power converters," *IEEE Transactions on Industrial Electronics*, vol. 62, no. 6, pp. 3409–3418, 2015.

[117] R. Ciric, A. Padilha, and L. Ochoa, "Power flow in four-wire distribution networks - general approach," *IEEE Transactions on Power Systems*, vol. 18, no. 4, pp. 893–893, 2005.

[118] P. Kotsampopoulos, D. Lagos, N. Hatziargyriou, M. O. Faruque, G. Lauss, O. Nzimako, P. Forsyth, M. Steurer, F. Ponci, A. Monti, V. Dinavahi, and K. Strunz, "A Benchmark System for Hardware-in-the-Loop Testing of Distributed Energy Resources," *IEEE Power and Energy Technology Systems Journal*, vol. 5, no. 3, pp. 94–103, 2018.

[119] R. Brandl, J. Montoya, T. Degner, and D. Strauss-Mincu, "Power system stability studies including real hardware using phasor power hardware-in-The-loop technology," in *2018 IEEE International Conference on Industrial Electronics for Sustainable Energy Systems, IESES 2018*, Hamilton, New Zealand, 2018, pp. 552–559.

[120] F. Ebe, B. Idlbi, D. E. Stakic, S. Chen, C. Kondzialka, M. Casel, G. Heilscher, C. Seitl, R. Bründlinger, and T. I. Strasser, "Comparison of power hardware-in-the-loop approaches for the testing of smart grid controls," *Energies*, vol. 11, no. 12, 2018.

[121] B. Mohandes, S. Acharya, M. S. El Moursi, A. Al-Sumaiti, H. Doukas, and S. Sgouridis, "Optimal Design of an Islanded Microgrid with Load Shifting Mechanism Between Electrical and Thermal Energy Storage Systems," *IEEE Transactions on Power Systems*, vol. 35, no. 4, pp. 2642–2657, 2020.

[122] C. Dufour, S. Cense, V. Jalili-Marandi, and J. Belanger, "Review of state-of-the-art solver solutions for HIL simulation of power systems, power electronic and motor drives," in *2013 15th European Conference on Power Electronics and Applications, EPE 2013*, Lille, France, 2013.

[123] CIGRE, "Technical Brochure 575: Benchmark Systems for Network Integration of Renewable and Distributed Energy Resources," Paris, France, 2014.

[124] S. J. Oh, C. H. Yoo, I. Y. Chung, and D. J. Won, "Hardware-in-the-Loop simulation of distributed intelligent energy management system for microgrids," *Energies*, vol. 6, no. 7, pp. 3263–3283, 2013.

[125] H. Fakham, A. Ahmidi, F. Colas, and X. Guillaud, "Multi-agent system for distributed voltage regulation of wind generators connected to distribution network," in *IEEE PES Innovative Smart Grid Technologies Conference Europe, ISGT Europe*, Gothenburg, Sweden, 2010.

[126] G. Pammer, "P-HIL Tutorial key design factors," 2019. [Online]. Available: https://erigrid.eu/wp-content/uploads/2019/11/07_FINAL_PHIL-Tutorial.pdf. [Accessed: 04-Apr-2020].

[127] F. Lehfuss, G. Lauss, P. Kotsampopoulos, N. Hatziargyriou, P. Crolla, and A. Roscoe, "Comparison of multiple power amplification types for power Hardware-in-the-Loop applications," in *2012 IEEE Workshop on Complexity in Engineering, COMPENG 2012*, Aachen, Germany, 2012, pp. 95–100.

[128] W. Ren, M. Steurer, and T. L. Baldwin, "An effective method for evaluating the accuracy of power hardware-in-the-loop simulations," *IEEE Transactions on Industry Applications*, vol. 45, no. 4, pp. 1484–1490, 2009.

[129] L. Dritschler, "Power-Hardware-in-the-Loop : Simulator-Hardware Interface Algorithms," Hochschule Karlsruhe - Technik und Wirtschaft, 2016.

[130] K. Ogata, *Modern Control Engineering*, Fifth Edit. Prentice Hal, 2010.

[131] G. Lauss, F. Lehfuss, B. Bletterie, T. Strasser, and R. Brundlinger, "Examination of LV grid phenomena by means of PHIL testing," in *Industrial Electronics Conference (IECON)*, Montreal, QC, Canada, 2012, pp. 4771–4776.

[132] A. Viehweider, G. Lauss, and L. Felix, "Stabilization of Power Hardware-in-the-Loop simulations of electric energy systems," *Simulation Modelling Practice and Theory*, vol. 19, no. 7, pp. 1699–1708, 2011.

[133] C. Mao, F. Leng, J. Li, S. Zhang, L. Zhang, R. Mo, D. Wang, J. Zeng, X. Chen, R. An, and Y. Zhao, "A 400-V/50-kVA Digital-Physical Hybrid Real-Time Simulation Platform for Power Systems," *IEEE Transactions on Industrial Electronics*, vol. 65, no. 5, pp. 3666–3676, 2018.

[134] G. Lauss and K. Strunz, "Multirate Partitioning Interface for Enhanced Stability of Power Hardware-in-the-Loop Real-Time Simulation," *IEEE Transactions on Industrial Electronics*, vol. 66, no. 1, pp. 595–605, 2019.

[135] M. Dargahi, A. Ghosh, and G. Ledwich, "Stability synthesis of power hardware-

in-the-loop (PHIL) simulation," in *IEEE Power and Energy Society General Meeting*, National Harbor, MD, USA, 2014.

[136] W. Ren, M. Steurer, and T. L. Baldwin, "Improve the stability and the accuracy of power hardware-in-the-loop simulation by selecting appropriate interface algorithms," *IEEE Transactions on Industry Applications*, vol. 44, no. 4, pp. 1286–1294, 2008.

[137] A. Riccobono, E. Liegmann, M. Pau, F. Ponci, and A. Monti, "Online Parametric Identification of Power Impedances to Improve Stability and Accuracy of Power Hardware-in-the-Loop Simulations," *IEEE Transactions on Instrumentation and Measurement*, vol. 66, no. 9, pp. 2247–2257, 2017.

[138] W. Ren, M. Steurer, and T. L. Baldwin, "Improve the stability and the accuracy of power hardware-in-the-loop simulation by selecting appropriate interface algorithms," *IEEE Transactions on Industry Applications*, vol. 44, no. 4, pp. 1286–1294, 2008.

[139] E. Guillo-Sansano, M. Syed, A. J. Roscoe, G. Burt, and F. Coffele, "Characterization of Time Delay in Power Hardware in the Loop Setups," *IEEE Transactions on Industrial Electronics*, vol. 68, no. 3, pp. 2703–2713, 2020.

[140] V. Kleftakis, A. Rigas, A. Vassilakis, P. Kotsampopoulos, and N. Hatziargyriou, "Power-Hardware-in-the-loop simulation of a D-STATCOM equipped MV network interfaced to an actual PV inverter," in *2013 4th IEEE/PES Innovative Smart Grid Technologies Europe (ISGT Europe 2013)*, Lyngby, Denmark, 2013.

[141] B. I. Craciun, T. Kerekes, D. Sera, R. Teodorescu, R. Brandl, T. Degner, D. Geibel, and H. Hernandez, "Grid integration of PV power based on PHIL testing using different interface algorithms," in *Industrial Electronics Conference (IECON)*, Vienna, Austria, 2013, pp. 5380–5385.

[142] Y. Zhou, H. Li, and L. Liu, "Integrated autonomous voltage regulation and islanding detection for high penetration PV applications," *IEEE Transactions on Power Electronics*, vol. 28, no. 6, pp. 2826–2841, 2013.

[143] F. Huerta, R. L. Tello, and M. Prodanovic, "Real-Time Power-Hardware-in-the-Loop Implementation of Variable-Speed Wind Turbines," *IEEE Transactions on Industrial Electronics*, vol. 64, no. 3, pp. 1893–1904, 2017.

[144] E. García-Martínez, J. Ballestín, J. Muñoz-Cruzado, and J. F. Sanz, "Analysis of a switched and linear power amplifier for Power Hardware-in-the-Loop testing of Smartgrid systems," in *IEEE International Conference on Emerging Technologies and Factory Automation (ETFA)*, Zaragoza, Spain, 2019.

[145] J. Langston, M. Steurer, M. Bosworth, D. Soto, D. Longo, M. Uva, and J. Carlton, "Power Hardware-in-the-Loop Simulation Testing of a 200 MJ Battery-Based Energy Magazine for Shipboard Applications," in *2019 IEEE Electric Ship*

Technologies Symposium (ESTS 2019), Washington, DC, USA, 2019, pp. 39–44.

[146] Z. Taylor, H. Akhavan-Hejazi, and H. Mohsenian-Rad, "Power hardware-in-loop simulation of grid-connected battery systems with reactive power control capability," in *2017 North American Power Symposium (NAPS 2017)*, Morgantown, WV, USA, 2017.

[147] S. Hubschneider, S. Kochanneck, B. Bohnet, M. Suriyah, I. Mauser, T. Leibfried, H. Schmeck, and M. Braun, "Requirements for Power Hardware-in-the-Loop Emulation of Distribution Grid Challenges," in *2018 53rd International Universities Power Engineering Conference, UPEC 2018*, Glasgow, UK, 2018.

[148] A. Chauhan and R. P. Saini, "A review on Integrated Renewable Energy System based power generation for stand-alone applications: Configurations, storage options, sizing methodologies and control," *Renewable and Sustainable Energy Reviews*, vol. 38, no. December, pp. 99–120, 2014.

[149] M. Zidar, P. S. Georgilakis, N. D. Hatziargyriou, T. Capuder, and D. Škrlec, "Review of energy storage allocation in power distribution networks: Applications, methods and future research," *IET Generation, Transmission and Distribution*, vol. 10, no. 3, pp. 645–652, 2016.

[150] C. K. Das, O. Bass, G. Kothapalli, T. S. Mahmoud, and D. Habibi, "Overview of energy storage systems in distribution networks: Placement, sizing, operation, and power quality," *Renewable and Sustainable Energy Reviews*, vol. 91, pp. 1205–1230, 2018.

[151] Y. Zheng, Z. Y. Dong, F. J. Luo, K. Meng, J. Qiu, and K. P. Wong, "Optimal allocation of energy storage system for risk mitigation of discos with high renewable penetrations," *IEEE Transactions on Power Systems*, vol. 29, no. 1, pp. 212–220, 2014.

[152] S. Wen, H. Lan, Q. Fu, D. C. Yu, and L. Zhang, "Economic allocation for energy storage system considering wind power distribution," *IEEE Transactions on Power Systems*, vol. 30, no. 2, pp. 644–652, 2015.

[153] P. Fortenbacher, A. Ulbig, and G. Andersson, "Optimal Placement and Sizing of Distributed Battery Storage in Low Voltage Grids Using Receding Horizon Control Strategies," *IEEE Transactions on Power Systems*, vol. 33, no. 3, pp. 2383–2394, 2018.

[154] Y. Zhang, S. Ren, Z. Y. Dong, Y. Xu, K. Meng, and Y. Zheng, "Optimal placement of battery energy storage in distribution networks considering conservation voltage reduction and stochastic load composition," *IET Generation, Transmission & Distribution*, vol. 11, no. 15, pp. 3862–3870, 2017.

[155] L. Meng, T. Dragicevic, and J. M. Guerrero, "Adaptive Control Design for Autonomous Operation of Multiple Energy Storage Systems in Power Smoothing

Applications," *IEEE Transactions on Industrial Electronics*, vol. 65, no. 8, pp. 6612–6624, 2018.

[156] R. Cárdenas, R. Peña, G. M. Asher, J. Clare, and R. Blasco-Giménez, "Control strategies for power smoothing using a flywheel driven by a sensorless vector-controlled induction machine operating in a wide speed range," *IEEE Transactions on Industrial Electronics*, vol. 51, no. 3, pp. 603–614, 2004.

[157] S. Shivashankar, S. Mekhilef, H. Mokhlis, and M. Karimi, "Mitigating methods of power fluctuation of photovoltaic (PV) sources - A review," *Renewable and Sustainable Energy Reviews*, vol. 59, pp. 1170–1184, 2016.

[158] R. C. Dugan, J. A. Taylor, and D. Montenegro, "Energy Storage Modeling for Distribution Planning," *IEEE Transactions on Industry Applications*, vol. 53, no. 2, pp. 954–962, 2016.

[159] "DIN EN 50160 - Merkmale der Spannung in öffentlichen Elektrizitätsversorgungsnetzen," Berlin, 2011.

[160] E. Demirok, D. Sera, R. Teodorescu, P. Rodriguez, and U. Borup, "Clustered PV inverters in LV networks: An overview of impacts and comparison of voltage control strategies," in *2009 IEEE Electrical Power and Energy Conference, EPEC 2009*, Montreal, QC, Canada, 2009.

[161] H. Nazaripouya, Y. Wang, P. Chu, H. R. Pota, and R. Gadh, "Optimal sizing and placement of battery energy storage in distribution system based on solar size for voltage regulation," in *IEEE Power and Energy Society General Meeting*, Denver, CO, USA, 2015.

[162] Y. Yang, H. Li, A. Aichhorn, J. Zheng, and M. Greenleaf, "Sizing strategy of distributed battery storage system with high penetration of photovoltaic for voltage regulation and peak load shaving," *IEEE Transactions on Smart Grid*, vol. 5, no. 2, pp. 982–991, 2014.

[163] O. Mansour, "Determining the Power and Energy Capacity of a Battery Energy Storage System Utilizing a Smoothing Feeder Profile to Accommodate High Photovoltaic Penetration on a Distribution Feeder," Portland State University, 2000.

[164] D. H. Popović, J. A. Greatbanks, M. Begović, and A. Pregelj, "Placement of distributed generators and reclosers for distribution network security and reliability," *International Journal of Electrical Power and Energy Systems*, vol. 27, no. 5–6, pp. 398–408, 2005.

[165] R. S. Al Abri, E. F. El-Saadany, and Y. M. Atwa, "Optimal placement and sizing method to improve the voltage stability margin in a distribution system using distributed generation," *IEEE Transactions on Power Systems*, vol. 28, no. 1, pp. 326–334, 2013.

[166] A. Giannitrapani, S. Paoletti, A. Vicino, and D. Zarrilli, "Optimal Allocation of Energy Storage Systems for Voltage Control in LV Distribution Networks," *IEEE Transactions on Smart Grid*, vol. 8, no. 6, pp. 2859–2870, 2017.

[167] T. Aziz, S. R. Deeba, and Nahid-Al-Masood, "Investigation of post-fault voltage recovery performance with battery-based energy storage system in a microgrid," in *2016 Australasian Universities Power Engineering Conference (AUPEC 2016)*, Brisbane, QLD, Australia, 2016.

[168] F. Ahmed, A. U. Mahin, A. Islam, and T. Aziz, "A comparative study on different placement algorithms of battery eanergy storage system in microgrid," in *4th International Conference on Advances in Electrical Engineering, ICAEE 2017*, Dhaka, Bangladesh, 2017, pp. 565–569.

[169] S. Hashemi and J. Østergaard, "Efficient Control of Energy Storage for Increasing the PV Hosting Capacity of LV Grids," *IEEE Transactions on Smart Grid*, vol. 9, no. 3, pp. 2295–2303, 2018.

[170] Y. Hase, *Handbook of Power Systems Engineering with Power Electronics Applications*, 2nd ed. John Wiley & Sons, 2012.

[171] K. Christakou, J. Y. Leboudec, M. Paolone, and D. C. Tomozei, "Efficient computation of sensitivity coefficients of node voltages and line currents in unbalanced radial electrical distribution networks," *IEEE Transactions on Smart Grid*, vol. 4, no. 2, pp. 741–750, 2013.

[172] C. Mugnier, K. Christakou, J. Jaton, M. De Vivo, M. Carpita, and M. Paolone, "Model-less/Measurement-based Computation of Voltage Sensitivities in Unbalanced Electrical Distribution Networks," in *19th Power Systems Computation Conference, PSCC 2016*, Genoa, Italy, 2016, pp. 1–7.

[173] K. M. I. U. Ranaweera, "Energy Storage for Control of Distributed Photovoltaic Power Systems," Norwegian University of Science and Technology (NTNU), 2019.

[174] S. Weckx, R. D'Hulst, and J. Driesen, "Voltage Sensitivity Analysis of a Laboratory Distribution Grid With Incomplete Data," *IEEE Transactions on Smart Grid*, vol. 6, no. 3, pp. 1271–1280, 2015.

[175] M. Vollmer, I. Rutter, and K. Böhm, "On complexity and efficiency of mutual information estimation on static and dynamic data," in *Advances in Database Technology - EDBT*, 2018, pp. 49–60.

[176] M. Vollmer, "Time-Efficient Analysis of Complex Dependencies," Karlsruhe Institute of Technology, 2020.

[177] H. Peng, "Feature Selection Based on Mutual Information: Criteria of Max-Dependency, Max-Relevance, and Min-Redundancy," *IEEE Transactions on*

Pattern Analysis and Machine Intelligence, vol. 27, no. 8, pp. 1226–1238, 2005.

[178] J. P. W. Pluim, J. B. A. A. Maintz, and M. A. Viergever, "Mutual-information-based registration of medical images: A survey," *IEEE Transactions on Medical Imaging*, vol. 22, no. 8, pp. 986–1004, 2003.

[179] A. Kraskov, H. Stögbauer, and P. Grassberger, "Estimating mutual information," *Physical Review E*, vol. 69, no. 6 2, 2004.

[180] S. Khan, S. Bandyopadhyay, A. R. Ganguly, S. Saigal, D. J. Erickson, V. Protopopescu, and G. Ostrouchov, "Relative performance of mutual information estimation methods for quantifying the dependence among short and noisy data," *Physical Review E*, vol. 76:026209, no. 2, 2007.

[181] S. Lu, S. Repo, D. Della Giustina, F. A. C. Figuerola, A. Lof, and M. Pikkarainen, "Real-Time Low Voltage Network Monitoring - ICT Architecture and Field Test Experience," *IEEE Transactions on Smart Grid*, vol. 6, no. 4, pp. 2002–2012, 2015.

[182] VDE, "VDE-AR-N 4105 Anwendungsregel:2018-11 Generators connected to the low-voltage distribution network: Technical requirements for the connection to and parallel operation with low-voltage distribution networks," 2018.

[183] H. Saboori, R. Hemmati, and M. A. Jirdehi, "Reliability improvement in radial electrical distribution network by optimal planning of energy storage systems," *Energy*, vol. 93, pp. 2299–2312, 2015.

[184] "Power Quality-Interface and Disturbance Recorder PQI-DA smart - A. Eberle GmbH & Co.KG." [Online]. Available: https://www.a-eberle.de/en/product-groups/pq-fix-installed/devices/pqi-da-smart. [Accessed: 06-Aug-2019].

[185] H. Maass, H. K. Cakmak, W. Suess, A. Quinte, W. Jakob, K. U. Stucky, and U. G. Kuehnapfel, "First evaluation results using the new electrical data recorder for power grid analysis," *IEEE Transactions on Instrumentation and Measurement*, vol. 62, no. 9, pp. 2384–2390, 2013.

[186] M. Hayn, V. Bertsch, and W. Fichtner, "Electricity load profiles in Europe: The importance of household segmentation," *Energy Research and Social Science*, vol. 3, pp. 30–45, 2014.

[187] T. Stetz, F. Marten, and M. Braun, "Improved low voltage grid-integration of photovoltaic systems in Germany," *IEEE Transactions on Sustainable Energy*, vol. 4, no. 2, pp. 534–542, 2013.

[188] A. Samadi, E. Shayesteh, R. Eriksson, B. Rawn, and L. Söder, "Coordinated Active Power-Dependent Voltage Regulation in Distribution Grids With PV Systems," *IEEE Transactions on Power Delivery*, vol. 71, no. 3, pp. 315–323, 2014.

[189] P. H. Divshali and L. Soder, "Improving PV Dynamic Hosting Capacity Using Adaptive Controller for STATCOMs," *IEEE Transactions on Energy Conversion*, vol. 34, no. 1, pp. 415–425, 2019.

[190] B. R. Ke, T. T. Ku, Y. L. Ke, C. Y. Chuang, and H. Z. Chen, "Sizing the battery energy storage system on a university campus with prediction of load and photovoltaic generation," *IEEE Transactions on Industrial Applications*, vol. 52, no. 2, pp. 1136–1147, 2015.

[191] Y. Ru, J. Kleissl, and S. Martinez, "Storage size determination for grid-connected photovoltaic systems," *IEEE Transactions on Sustainable Energy*, vol. 4, no. 1, pp. 68–81, 2013.

[192] S. Majumder, S. A. Khaparde, A. P. Agalgaonkar, P. Ciufo, S. Perera, and S. V. Kulkarni, "DFT-Based Sizing of Battery Storage Devices to Determine Day-Ahead Minimum Variability Injection Dispatch with Renewable Energy Resources," *IEEE Transactions on Smart Grid*, vol. 10, no. 1, pp. 626–638, 2019.

[193] P. Fortenbacher, M. Zellner, and G. Andersson, "Optimal sizing and placement of distributed storage in low voltage networks," in *19th Power Systems Computation Conference, PSCC 2016*, Genoa, Italy, 2016.

[194] N. Nguyen-Hong, H. Nguyen-Duc, and Y. Nakanishi, "Optimal Sizing of Energy Storage Devices in Isolated Wind-Diesel Systems Considering Load Growth Uncertainty," *IEEE Transactions on Industry Applications*, vol. 54, no. 3, pp. 1983–1991, 2018.

[195] K. Baker, G. Hug, and X. Li, "Optimal storage sizing using two-stage stochastic optimization for intra-hourly dispatch," in *2014 North American Power Symposium, NAPS 2014*, Pullman, WA, USA, 2014.

[196] A. Oudalov, R. Cherkaoui, and A. Beguin, "Sizing and optimal operation of battery energy storage system for peak shaving application," in *2007 IEEE Lausanne PowerTech*, Lausanne, Switzerland, 2007, pp. 621–625.

[197] J. Xiao, L. Bai, F. Li, H. Liang, and C. Wang, "Sizing of energy storage and diesel generators in an isolated microgrid Using Discrete Fourier Transform (DFT)," *IEEE Transactions on Sustainable Energy*, vol. 5, no. 3, pp. 907–916, 2014.

[198] Y. Xiang, W. Han, J. Zhang, J. Liu, and Y. Liu, "Optimal Sizing of Energy Storage System in Active Distribution Networks Using Fourier-Legendre Series Based State of Energy Function," *IEEE Transactions on Power Systems*, vol. 33, no. 2, pp. 2313–2315, 2018.

[199] A. Nagarajan and R. Ayyanar, "Design and Strategy for the Deployment of Energy Storage Systems in a Distribution Feeder With Penetration of Renewable Resources," *IEEE Transactions on Sustainable Energy*, vol. 6, no. 3, pp. 1085–1092, 2015.

[200] M. Nick, R. Cherkaoui, and M. Paolone, "Optimal allocation of dispersed energy storage systems in active distribution networks for energy balance and grid support," *IEEE Transactions on Power Systems*, vol. 29, no. 5, pp. 2300–2310, 2014.

[201] D. D. Sharma, S. N. Singh, B. S. Rajpurohit, and F. G. Longatt, "Critical load profile estimation for sizing of battery storage system," in *IEEE Power and Energy Society General Meeting 2015*, Denver, CO, USA, 2015, pp. 9–13.

[202] P. D. Brown, J. A. Peças Lopes, and M. A. Matos, "Optimization of pumped storage capacity in an isolated power system with large renewable penetration," *IEEE Transactions on Power Systems*, vol. 23, no. 2, pp. 523–531, 2008.

[203] F. Kazhamiaka, Y. Ghiassi-Farrokhfal, S. Keshav, and C. Rosenberg, "Comparison of Different Approaches for Solar PV and Storage Sizing," *IEEE Transactions on Sustainable Computing (Early Access)*, 2019.

[204] J. Buhler and M. Tompa, "Finding motifs using random projections," *Journal of Computational Biology*, vol. 9, no. 2, pp. 225–242, 2002.

[205] N. Ludwig, S. Waczowicz, R. Mikut, and V. Hagenmeyer, "Assessment of unsupervised standard pattern recognition methods for industrial energy time series," in *e-Energy 2018 - Proceedings of the 9th ACM International Conference on Future Energy Systems*, Karlsruhe, Germany, 2018, pp. 434–435.

[206] C. A. Hill, M. C. Such, D. Chen, J. Gonzalez, and W. M. K. Grady, "Battery energy storage for enabling integration of distributed solar power generation," *IEEE Transactions on Smart Grid*, vol. 3, no. 2, pp. 850–857, 2012.

[207] N. Ludwig, "Data-Driven Methods for Demand-Side Flexibility in Energy Systems," Karlsruhe Institute of Technology (KIT), 2020.

[208] C. Cassisi, M. Aliotta, A. Cannata, P. Montalto, D. Patanè, A. Pulvirenti, and L. Spampinato, "Motif Discovery on Seismic Amplitude Time Series: The Case Study of Mt Etna 2011 Eruptive Activity," *Pure and Applied Geophysics*, vol. 170, no. 4, pp. 529–545, 2013.

[209] E. F. Gomes, A. M. Jorge, and P. J. Azevedo, "Classifying heart sounds using multiresolution time series motifs: An exploratory study," in *Proceedings of the International C* Conference on Computer Science and Software Engineering*, Porto, Portugal, 2013, pp. 23–30.

[210] N. Ludwig, L. Barth, D. Wagner, and V. Hagenmeyer, "Industrial demand-side flexibility: A benchmark data set," in *e-Energy 2019 - Proceedings of the 10th ACM International Conference on Future Energy Systems*, Phoenix, AZ, USA, 2019, pp. 460–473.

[211] Y. Simmhan and M. U. Noor, "Scalable prediction of energy consumption using

incremental time series clustering," in *2013 IEEE International Conference on Big Data, Big Data 2013*, Silicon Valley, CA, USA, 2013, pp. 29–36.

[212] "Powerwall | Tesla Heimspeicher." [Online]. Available: https://www.tesla.com/de_DE/powerwall. [Accessed: 19-Aug-2020].

[213] J. Liu, Y. Miura, H. Bevrani, and T. Ise, "Enhanced Virtual Synchronous Generator Control for Parallel Inverters in Microgrids," *IEEE Transactions on Smart Grid*, vol. 8, no. 5, pp. 2268–2277, 2017.

[214] D. B. W. Abeywardana, B. Hredzak, V. G. Agelidis, and G. D. Demetriades, "Supercapacitor sizing method for energy-controlled filter-based hybrid energy storage systems," *IEEE Transactions on Power Electronics*, vol. 32, no. 2, pp. 1626–1637, 2017.

[215] I. N. Moghaddam and B. Chowdhury, "Optimal sizing of Hybrid Energy Storage Systems to mitigate wind power fluctuations," in *IEEE Power and Energy Society General Meeting*, Boston, MA, USA, 2016.

[216] Y. V. Makarov, P. Du, M. C. W. Kintner-Meyer, C. Jin, and H. F. Illian, "Sizing energy storage to accommodate high penetration of variable energy resources," *IEEE Transactions on Sustainable Energy*, vol. 3, no. 1, pp. 34–40, 2012.

[217] L. F. Pak and V. Dinavahi, "Real-time simulation of a wind energy system based on the doubly-fed induction generator," *IEEE Transactions on Power Systems*, vol. 24, no. 3, pp. 1301–1309, 2009.

[218] M. Park and I. K. Yu, "A novel real-time simulation technique of photovoltaic generation systems using RTDS," *IEEE Transactions on Energy Conversion*, vol. 19, no. 1, pp. 164–169, 2004.

[219] J. Langston, M. Steurer, S. Woodruff, T. Baldwin, and J. Tang, "A generic real-time computer simulation model for superconducting fault current limiters and its application in system protection studies," *IEEE Transactions on Applied Superconductivity*, vol. 15, no. 2, pp. 2090–2093, 2005.

[220] J. Geisbuesch, S. Karrari, P. Kreideweis, W. T. Batista De Sousa, and M. Noe, "Set-Up of a Dynamic Multi-Purpose Power-Hardware-in-The-Loop System for New Technologies Integration," in *2018 IEEE Workshop on Complexity in Engineering, COMPENG 2018*, Florence, Italy, 2018.

[221] A. Collet, J. C. Crebier, and A. Chureau, "Multi-cell battery emulator for advanced battery management system benchmarking," in *2011 IEEE International Symposium on Industrial Electronics (ISIE 2011)*, Gdansk, Poland, 2011, pp. 1093–1099.

[222] C. Seitl, C. Messner, H. Popp, and J. Kathan, "Emulation of a high voltage home storage battery system using a power hardware-in-the-loop approach," in

Industrial Electronics Conference (IECON), Florence, Italy, 2016.

[223] J. D. Boles, Y. Ma, J. Wang, D. Osipov, L. M. Tolbert, and F. Wang, “Converter-based emulation of battery energy storage systems (BESS) for grid applications,” *IEEE Transactions on Industry Applications*, vol. 55, no. 4, pp. 4020–4032, 2019.

[224] I. Munteanu, A. I. Bratcu, S. Bacha, D. Roye, and J. Guiraud, “Hardware-in-the-loop-based simulator for a class of variable-speed wind energy conversion systems: Design and performance assessment,” in *IEEE Transactions on Energy Conversion*, 2010, vol. 25, no. 2, pp. 564–576.

[225] R. Ayop and C. W. Tan, “Rapid prototyping of photovoltaic emulator using buck converter based on fast convergence resistance feedback method,” *IEEE Transactions on Power Electronics*, vol. 34, no. 9, pp. 8715–8723, 2019.

[226] A. F. Ebrahim, S. M. W. Ahmed, S. E. Elmasry, and O. A. Mohammed, “Implementation of a PV emulator using programmable DC power supply,” in *IEEE SoutheastCon 2015*, Fort Lauderdale, FL, USA, 2015.

[227] A. Schmitt, J. Richter, M. Gommeringer, T. Wersal, and M. Braun, “A novel 100 kW power hardware-in-the-loop emulation test bench for permanent magnet synchronous machines with nonlinear magnetics,” in *8th IET International Conference on Power Electronics, Machines and Drives (PEMD 2016)*, Glasgow, UK, 2016.

[228] V. T. Tran, M. R. Islam, D. Sutanto, and K. M. Muttaqi, “Mitigation of Solar PV Intermittency Using Ramp-Rate Control of Energy Buffer Unit,” *IEEE Transactions on Energy Conversion*, vol. 34, no. 1, pp. 435–445, 2019.

[229] S. A. Khajehoddin, M. Karimi-Ghartemani, and M. Ebrahimi, “Grid-supporting inverters with improved dynamics,” *IEEE Transactions on Industrial Electronics*, vol. 66, no. 5, pp. 3655–3667, 2019.

[230] R. Sheikh, *Electric Storage*. Rijeka, Croatia: Sciyo, 2010.

[231] R. Krishnan, *Permanent Magnet Synchronous and Brushless DC Motor Drives*. Boca Raton: CRC Press, 2010.

[232] A. Schmitt, “Hochdynamische Power Hardware-in-the-Loop Emulation hoch ausgenutzter Synchronmaschinen mit einem Modularen-Multiphasen-Multilevel Umrichter,” Karlsruhe Institute of Technology (KIT), 2017.

[233] N. P. Quang and J.-A. Dittrich, *Vector Control of Three-Phase AC Machines*. 2008.

[234] T. Gemaßmer, J. Richter, M. Schnarrenberger, and M. Braun, “High dynamic rotor oriented current control for permanent magnet synchronous machines w saturation characteristics,” in *PCIM Europe Conference Proceedings*, Nure

Germany, 2014, pp. 153–160.

[235] L. Wang, S. Chai, D. Yoo, L. Gan, and K. Ng, *c*. John Wiley & Sons, 2014.

[236] D. Schröder, *Elektrische Antriebe - Regelung von Antriebssystemen*. Berlin: Springer, 2015.

[237] H. Akagi, E. H. Watanabe, and M. Aredes, *Instantaneous Power Theory and Applications to Power Conditioning*. John Wiley & Sons, Inc, 2017.

[238] J. Richter, "Modellbildung, Parameteridentifikation und Regelung hoch ausgenutzter Synchronmaschinen," Karlsruhe Institute of Technology (KIT), 2016.

[239] K. Patil and B. Mehta, "Modeling and simulation of variable speed wind turbine with direct drive permanent magnet synchronous generator," University of Tasmania, 2014.

[240] CIGRE, "Technical Brochure 604: Guide for the development of models for HVDC," 2014.

[241] S. Bacha, I. Munteanu, and A. I. Bratcu, *Power Electronic Converters Modeling and Control*. Springer, 2014.

[242] J. Allmeling and N. Felderer, "Sub-cycle average models with integrated diodes for real-time simulation of power converters," in *2017 IEEE Southern Power Electronics Conference, SPEC 2017*, Puerto Varas, Chile, 2017.

[243] A. Yazdani and R. Iravani, *Voltage-Sourced Converters in Power Systems: Modeling, Control, and Applications*. John Wiley & Sons, Inc., 2010.

[244] H. Abu-Rub, M. Malinowski, and K. Al-Haddad, *Power Electronics for Renewable Energy Systems, Transportation and Industrial Applications*. West Sussex: Wiley-IEEE Press, 2014.

[245] S. Abourida, J. Bélanger, and C. Dufour, "Real-time HIL simulation of a complete PMSM drive at 10 μs time step," in *2005 European Conference on Power Electronics and Applications*, Niigata, Japan, 2005.

[246] G. Sybille and H. Le-Huy, "A comparative study on real-time simulation methods f... er converters," in *2006 IEEE International Symposium on Industrial* ... ntreal, QC, Canada, 2006.

... M. Laursen, and S. Hansen, "Voltage vector based control for ... y applications," in *IEEE International Symposium on Industrial* ... pp. 3845–3850.

... S. K. Sul, and K. Ide, "Maximum torque per ampere (MTPA) control of an IPM machine based on signal injection considering inductance saturation," *IEEE Transactions on Power Electronics*, vol. 28, no. 1, pp. 488–497, 2013.

[249] Q. Wang, H. Yuan, X. Jiang, and X. Xing, "A simple sensorless control strategy for IPMSM in 1 MW flywheel energy storage system," in *2016 IEEE 8th International Power Electronics and Motion Control Conference, IPEMC-ECCE Asia 2016*, Hefei, China, 2016, pp. 1923–1928.

[250] B. H. Kenny, P. E. Kascak, R. Jansen, T. Dever, and W. Santiago, "Control of a high-speed flywheel system for energy storage in space applications," *IEEE Transactions on Industry Applications*, vol. 41, no. 4, pp. 1029–1038, 2005.

[251] R. Teodorescu, M. Liserre, and P. Rodriıguez, *Grid Converters for Photovoltaic and Wind Power Systems*. John Wiley & Sons Ltd, 2011.

[252] H. H. Abdeltawab and Y. A. R. I. Mohamed, "Robust Energy Management of a Hybrid Wind and Flywheel Energy Storage System Considering Flywheel Power Losses Minimization and Grid-Code Constraints," *IEEE Transactions on Industrial Electronics*, vol. 63, no. 7, pp. 4242–4254, 2016.

[253] C. Mi, G. R. Slemon, and R. Bonert, "Modeling of iron losses of permanent-magnet synchronous motors," *IEEE Transactions on Industry Applications*, vol. 39, no. 3, pp. 734–742, 2003.

[254] A. Baggini, *Handbook of Power Quality*. West Sussex: John Wiley & Sons Ltd, 2008.

[255] F. Diaz-Gonzalez, A. Sumper, O. Gomis-Bellmunt, and R. Villafafila-Robles, "Modeling and validation of a flywheel energy storage lab-setup," in *IEEE PES Innovative Smart Grid Technologies Conference Europe*, Berlin, Germany, 2012.

[256] G. F. Lauss, M. O. Faruque, K. Schoder, C. Dufour, A. Viehweider, and J. Langston, "Characteristics and design of power hardware-in-The-loop simulations for electrical power systems," *IEEE Transactions on Industrial Electronics*, vol. 63, no. 1, pp. 406–417, 2016.

[257] VDE, "DIN VDE V 0124-100:2020-06: Grid integration of generator plants Low-voltage – Test requirements for generator units to be connected to and operated in parallel with low-voltage distribution networks," 2020.

[258] National Grid ESO, "Technical Report on the events of 9 August 2019," 2019.

[259] "Historic frequency data | National Grid ESO." [Online]. Available: https://www.nationalgrideso.com/balancing-services/frequency-response-services/historic-frequency-data. [Accessed: 08-Nov-2020].

[260] K. L. Lo, L. J. Peng, J. F. Macqueen, A. O. Ekwue, and D. T. Y. Cheng, "An

extended Ward equivalent approach for power system security assessment," *Electric Power Systems Research*, vol. 42, no. 3, pp. 181–188, 1997.

[261] S. I. Nanou, A. G. Papakonstantinou, and S. A. Papathanassiou, "A generic model of two-stage grid-connected PV systems with primary frequency response and inertia emulation," *Electric Power Systems Research*, vol. 127, pp. 186–196, 2015.

[262] "SOLARWATT Vision 60M Style - Solarmodule mit attraktiven Design." [Online]. Available: https://www.solarwatt.de/solarmodule/glas-glas/vision-60m-style. [Accessed: 12-Oct-2020].

[263] Australian Energy Market Operator, "Black system, South Australia, 28 September 2016," 2017.

[264] C. Pradhan, C. N. Bhende, and A. K. Samanta, "Adaptive virtual inertia-based frequency regulation in wind power systems," *Renewable Energy*, vol. 115, pp. 558–574, 2018.

[265] U. Markovic, N. Früh, P. Aristidou, and G. Hug, "Interval-based adaptive inertia and damping control of a virtual synchronous machine," in *2019 IEEE Milan PowerTech*, Milan, Italy, 2019.

[266] IEEE PES Power System Dynamic Performance Committee, "Dynamic Models for Turbine-Governors in Power System Studies," 2013.

[267] H. Bevrani, *Robust Power System Frequency Control*, 2nd ed. Springer, 2009.

[268] J. Alipoor, Y. Miura, and T. Ise, "Power system stabilization using virtual synchronous generator with alternating moment of inertia," *IEEE Journal of Emerging and Selected Topics in Power Electronics*, vol. 3, no. 2, pp. 451–458, 2015.

[269] F. Wang, L. Zhang, X. Feng, and H. Guo, "An Adaptive Control Strategy for Virtual Synchronous Generator," *IEEE Transactions on Industry Applications*, vol. 54, no. 5, pp. 5124–5133, 2018.

[270] D. Li, Q. Zhu, S. Lin, and X. Y. Bian, "A Self-Adaptive Inertia and Damping Combination Control of VSG to Support Frequency Stability," *IEEE Transactions on Energy Conversion*, vol. 32, no. 1, pp. 397–398, 2017.

[271] X. Hou, Y. Sun, X. Zhang, J. Lu, P. Wang, and J. M. Guerrero, "Improvement of Frequency Regulation in VSG-Based AC Microgrid Via Adaptive Virtual Inertia," *IEEE Transactions on Power Electronics*, vol. 35, no. 2, pp. 1589–1602, 2020.

[272] J. Morren, S. W. H. de Haan, W. L. Kling, and J. A. Ferreira, "Wind turbines emulating inertia and supporting primary frequency control," *IEEE Transactions on Power Systems*, vol. 21, no. 1, pp. 433–434, 2006.

[273] V. Karapanos, S. De Haan, and K. Zwetsloot, "Real time simulation of a power system with VSG hardware in the loop," in *IECON Proceedings (Industrial Electronics Conference)*, Melbourne, VIC, Australia, 2011, pp. 3748–3754.

[274] S. D'Arco, J. A. Suul, and O. B. Fosso, "A Virtual Synchronous Machine implementation for distributed control of power converters in SmartGrids," *Electric Power Systems Research*, vol. 122, pp. 180–197, 2015.

[275] J. Liu, Y. Miura, and T. Ise, "Power Quality improvement of microgrids by virtual synchronous generator control," in *2016 Electric Power Quality and Supply Reliability, PQ 2016*, 2016, pp. 119–124.

[276] H. Zhao, Q. Yang, and H. Zeng, "Multi-loop virtual synchronous generator control of inverter-based DGs under microgrid dynamics," *IET Generation, Transmission and Distribution*, vol. 11, no. 3, pp. 795–803, 2017.

[277] J. Van De Vyver, J. D. M. De Kooning, B. Meersman, L. Vandevelde, and T. L. Vandoorn, "Droop Control as an Alternative Inertial Response Strategy for the Synthetic Inertia on Wind Turbines," *IEEE Transactions on Power Systems*, vol. 31, no. 2, pp. 1129–1138, 2016.

[278] Y. Fu, Y. Wang, and X. Zhang, "Integrated wind turbine controller with virtual inertia and primary frequency responses for grid dynamic frequency support," *IET Renewable Power Generation*, vol. 11, no. 8, pp. 1129–1137, 2017.

[279] J. Hu, S. Wang, W. Tang, and X. Xiong, "Full-capacity wind turbine with inertial support by adjusting phase-locked loop response," *IET Renewable Power Generation*, vol. 11, no. 1, pp. 44–53, 2017.

[280] A. Hosseinipour and H. Hojabri, "Virtual inertia control of PV systems for dynamic performance and damping enhancement of DC microgrids with constant power loads," *IET Renewable Power Generation*, vol. 12, no. 4, pp. 430–438, 2018.

[281] J. Khazaei, Z. Tu, and W. Liu, "Small-Signal Modeling and Analysis of Virtual Inertia-Based PV Systems," *IEEE Transactions on Energy Conversion*, vol. 35, no. 2, pp. 1129–1138, 2020.

[282] U. Tamrakar, D. Shrestha, M. Maharjan, B. P. Bhattarai, T. M. Hansen, and R. Tonkoski, "Virtual inertia: Current trends and future directions," *Applied Sciences*, vol. 7, no. 7, p. 654, 2017.

[283] F. Milano, F. Dorfler, G. Hug, D. J. Hill, and G. Verbič, "Foundations and challenges of low-inertia systems," in *20th Power Systems Computation Conference, PSCC 2018*, Dublin, Ireland, 2018.

[284] J. Fang, H. Li, Y. Tang, and F. Blaabjerg, "Distributed Power System Virtual Inertia Implemented by Grid-Connected Power Converters," *IEEE Transactions*

on Power Electronics, vol. 33, no. 10, pp. 8488–8499, 2018.

[285] M. Jozef, D. Ioan, V. Knap, R. Sinha, M. Swierczynski, D. Stroe, and S. Chaudhary, "Grid Inertial Response with Lithium-ion Battery Energy Storage Systems," in *23rd IEEE International Symposium on Industrial Electronics, ISIE 2014*, 2014.

[286] S. M. Alhejaj and F. M. Gonzalez-Longatt, "Impact of inertia emulation control of grid-scale BESS on power system frequency response," in *2016 International Conference for Students on Applied Engineering, ICSAE 2016*, Newcastle Upon Tyne, UK, 2017, pp. 254–258.

[287] J. Fang, H. Li, Y. Tang, and F. Blaabjerg, "On the Inertia of Future More-Electronics Power Systems," *IEEE Journal of Emerging and Selected Topics in Power Electronics*, vol. 7, no. 4, pp. 2130–2146, 2019.

[288] J. Fang, Y. Tang, H. Li, and X. Li, "A Battery/Ultracapacitor Hybrid Energy Storage System for Implementing the Power Management of Virtual Synchronous Generators," *IEEE Transactions on Power Electronics*, vol. 33, no. 4, pp. 2820–2824, 2018.

[289] Q. C. Zhong and G. Weiss, "Synchronverters: Inverters that mimic synchronous generators," *IEEE Transactions on Industrial Electronics*, vol. 58, no. 4, pp. 1259–1267, 2011.

[290] H. A. Alsiraji and R. El-Shatshat, "Comprehensive assessment of virtual synchronous machine based voltage source converter controllers," *IET Generation, Transmission and Distribution*, vol. 11, no. 7, pp. 1762–1769, 2017.

[291] Q. C. Zhong, P. L. Nguyen, Z. Ma, and W. Sheng, "Self-synchronized synchronverters: Inverters without a dedicated synchronization unit," *IEEE Transactions on Power Electronics*, vol. 29, no. 2, pp. 617–630, 2014.

[292] S. D'Arco and J. A. Suul, "Virtual synchronous machines - Classification of implementations and analysis of equivalence to droop controllers for microgrids," in *2013 IEEE Grenoble PowerTech*, Grenoble, France, 2013, pp. 1–7.

[293] V. Karapanos, S. W. . De Haan, and K. Zwetsloot, "Testing a Virtual Synchronous Generator in a Real Time Simulated Power System," in *International Conference on Power Systems Transients IPST 2011*, Delft, Netherlands, 2011.

[294] E. Rakhshani and P. Rodriguez, "Inertia Emulation in AC / DC Interconnected Power," *IEEE Transactions on Power Systems*, vol. 32, no. 5, pp. 3338–3351, 2017.

[295] S. Yang, J. Fang, Y. Tang, H. Qiu, C. Dong, and P. Wang, "Modular Multilevel Converter Synthetic Inertia-Based Frequency Support for Medium-Voltage Microgrids," *IEEE Transactions on Industrial Electronics*, vol. 66, no. 11, pp.

8992–9002, 2019.

[296] M. A. Torres L., L. A. C. Lopes, L. A. Morán T., and J. R. Espinoza C., "Self-tuning virtual synchronous machine: A control strategy for energy storage systems to support dynamic frequency control," *IEEE Transactions on Energy Conversion*, vol. 29, no. 4, pp. 833–840, 2014.

[297] B. Rathore, S. Chakrabarti, and S. Anand, "Frequency response improvement in microgrid using optimized VSG control," in *2016 National Power Systems Conference, NPSC 2016*, Bhubaneswar, India, 2017, pp. 1–6.

[298] U. Markovic, Z. Chu, P. Aristidou, and G. Hug, "LQR-Based Adaptive Virtual Synchronous Machine for Power Systems With High Inverter Penetration," *IEEE Transactions on Sustainable Energy*, vol. 10, no. 3, pp. 1501–1512, 2019.

[299] J. Alipoor, Y. Miura, and T. Ise, "Stability assessment and optimization methods for microgrid with multiple VSG units," *IEEE Transactions on Smart Grid*, vol. 9, no. 2, pp. 1462–1471, 2018.

[300] A. Karimi, Y. Khayat, M. Naderi, T. Dragičević, R. Mirzaei, F. Blaabjerg, and H. Bevrani, "Inertia Response Improvement in AC Microgrids: A Fuzzy-Based Virtual Synchronous Generator Control," *IEEE Transactions on Power Electronics*, vol. 35, no. 4, pp. 4321–4331, 2020.

[301] L. Zhang, L. Harnefors, and H. P. Nee, "Power-synchronization control of grid-connected voltage-source converters," *IEEE Transactions on Power Systems*, vol. 25, no. 2, pp. 809–820, 2010.

[302] J. Fang, R. Zhang, H. Li, and Y. Tang, "Frequency Derivative-based Inertia Enhancement by Grid-Connected Power Converters with a Frequency-Locked-Loop," *IEEE Transactions on Smart Grid*, vol. 10, no. 5, pp. 4918–4927, 2018.

[303] Q. Peng, Q. Jiang, Y. Yang, T. Liu, H. Wang, and F. Blaabjerg, "On the Stability of Power Electronics-Dominated Systems: Challenges and Potential Solutions," *IEEE Transactions on Industry Applications*, vol. 55, no. 6, pp. 7657–7670, 2019.

[304] P. Rodríguez, A. Luna, R. S. Muñoz-Aguilar, I. Etxeberria-Otadui, R. Teodorescu, and F. Blaabjerg, "A stationary reference frame grid synchronization system for three-phase grid-connected power converters under adverse grid conditions," *IEEE Transactions on Power Electronics*, vol. 27, no. 1, pp. 99–112, 2012.

[305] S. Papathanassiou, H. Nikos, and S. Kai, "A benchmark low voltage microgrid network," in *Proceedings of the CIGRE symposium: power systems with dispersed generation*, Athens, Greece, 2005.

[306] K. Soumalas, G. Messinis, and N. Hatziargyriou, "A data driven approach to distribution network topology identification," in *2017 IEEE Manchester PowerTech*, Manchester, UK, 2017.

[307] IEC, "IEC 61000-2-2: Electromagnetic compatibility (EMC) - Part 2-2: Environment - Compatibility levels for low-frequency conducted disturbances and signalling in public low-voltage power supply systems," 2002.

Publications

Main publications:

1. **S. Karrari**, N. Ludwig, G. De Carne and M. Noe, "Sizing of Hybrid Energy Storage Systems Using Recurring Daily Patterns," in IEEE Transactions on Smart Grid, vol. 13, no. 4, pp. 3290-3300, July 2022.
2. **S. Karrari**, G. De Carne, and M. Noe, "Model validation of a high-speed flywheel energy storage system using power hardware-in-the-loop testing," Journal of Energy Storage, vol. 43, no. 103177, 2021.
3. **Karrari,** H. R. Baghaee, G. De Carne, M. Noe, and J. Geisbüsch, "Adaptive Inertia Emulation Control for High-speed Flywheel Energy Storage Systems," IET Generation, Transmission & Distribution, vol. 14, no. 22, pp. 5047–5059, 2020.
4. **S. Karrari**, M. Vollmer, G. De Carne, K. Böhm, J. Geisbüsch, "A Data-driven Approach for Estimating Relative Voltage Sensitivity," in the 2020 IEEE Power & Energy Society General Meeting, Montreal, 2020.
5. **S. Karrari**, N. Ludwig, M. Noe, V. Hagenmeyer, "Sizing Centralized Energy Storage Systems in Distribution Networks Using Standard Patterns," in IEEE PowerTech 2019, Milan, 2019.
6. **S. Karrari**, M. Noe, and J. Geisbüsch, "High-speed Flywheel Energy Storage System (FESS) for Voltage and Frequency Support in Low Voltage Distribution Networks," in 3rd IEEE International Conference on Intelligent Energy and Power Systems (IEPS 2018), Kharkov, 2018, pp. 176–182.
7. **S. Karrari**, M. Noe, and J. Geisbüsch, "Real-time simulation of high-speed Flywheel Energy Storage System (FESS) for distribution networks," in the 9th ACM International Conference on Future Energy Systems (e-Energy 2018), Karlsruhe, 2018.
8. J. Geisbüsch, **S. Karrari**, P. Kreideweis, W. Tiago, B. De Sousa, and M. Noe, "Set-up of a Dynamic Multi-purpose Power-Hardware- in-the-Loop System for New Technologies Integration," in the 2018 IEEE Workshop on Complexity in Engineering (COMPENG 2018), Florence, 2018.

Other Contributions:

9. V. Arzamasov, R. Schwerdt, **S. Karrari**, K. Böhm, T. B. Nguyen, "Privacy Measures and Storage Technologies for Battery-Based Load Hiding—an Overview and Experimental Study," in Proceedings of the 11th ACM

International Conference on Future Energy Systems (e-Energy 2020), Melbourne, 2020.

10. M. Vollmer, H. Trittenbach, **S. Karrari**, A. Englhardt, P. Bielski, K. Böhm, "Energy Time-Series Features for Emerging Applications on the Basis of Human-Readable Machine Descriptions," in 2nd International Workshop on Energy Data and Analytics, Phoenix, 2019.
11. W. T. B. De Sousa, D. Kottonau, **S. Karrari**, J. Geisbusch, and M. Noe, "Deployment of a Resistive Superconducting Fault Current Limiter for Improvement of Voltage Quality and Transient Recovery Voltage," IEEE Transactions on Applied Superconductivity, vol. 31, no. 1, 2021.

Conference Poster and Oral Presentations:

1. **S. Karrari**, M. Noe, J. Geisbüsch, "High-speed Flywheel Energy Storage System (FESS) for Voltage and Frequency Support in Low-voltage Distribution Networks," in International Conference on Integration of Renewable and Distributed Energy Resources (IRED 2018), Vienna, 2018.
2. **S. Karrari**, M. Noe, J. Geisbüsch, "Voltage and Frequency Support in Low Voltage Distribution Networks by a High-speed Flywheel Energy Storage System (FESS)," in 10th International Conference on Real-Time Simulation (RT18), Paris, 2018.
3. **S. Karrari**, M. Noe, J. Geisbüsch, "High-speed Flywheel Energy Storage Systems for Voltage Support in Low Voltage Networks," WE-Heraeus-Seminar on Nonlinear Dynamics, Optimization and Control of Distributed Energy Systems, Bad Honnef, 2018
4. **S. Karrari**, M. Noe, J. Geisbüsch, "Real-time simulation and PHIL testing: Flywheel Energy Storage Systems," in Workshop Echtzeitsimulation in der Energietechnik, Karlsruhe, 2018.
5. **S. Karrari**, M. Noe, J. Geisbüsch, "Integration of Energy Storage Systems into Distribution Grids: A High-Speed Flywheel Energy Storage System," in International Workshop on Dynamics in Power-to-X-Systems, Karlsruhe, 2018.

Appendix A. A Summary of Power Quality Measurements

As described in subsection 4.1.4.2, in this work, measurement data have been collected from several low voltage distribution grids in southern Germany, which were used for the purpose of allocation and sizing of the FESS using data-driven methods. These data were recorded using A. Eberle's "PQI-DA smart" device, which is a power quality and disturbance recorder. In this appendix, after giving a short summary of the power quality requirements for the low voltage distribution grid in Germany, the status of the power quality at the measured low voltage grids are summarized.

In short, at none of the measured locations, a major power quality violation was observed, and a high power quality is being delivered by the distribution network operator at the investigated locations. However, several interesting incidents and observations were recorded, which are presented in this appendix.

A.1 Power Quality Requirements for Low Voltage Grids in Germany

This section summarizes the minimum power quality requirements that the system operator should provide for the private and industrial customers, connected to the low voltage networks. These requirements include the maximum variation in the voltage amplitude, frequency, harmonics, flicker, and control signal level. These requirements are provided in the grid codes DIN EN 50160 and DIN EN 61000-2-2 for public grids (or DIN EN 61000-2-4 for industrial grids).

The objective of the system operator is to provide a balanced three-phase sinusoidal voltage with 50 Hz frequency. However, the variations in load, generation, switching of devices and lines, nonlinear loads, transformers during energizing, and faults may lead to deviation of these properties from their ideal values. Nevertheless, grid codes obligate the network operators to keep the electrical characteristics of the provided electricity supply within some predefined limits. Table A.1 summarizes the grid code requirements of the electricity supply in the low voltage networks in Germany. More details about each specific requirement can be found in DIN EN 50160.

Table A.1: Summary of the power quality requirements at the low voltage level in Germany [159], [307].

Electricity Supply Characteristic	**Acceptable Range/Values**	**Evaluation Criteria**	**Evaluation Period**
Voltage RMS	±10 %	95 % of the 10-minute averages of voltage RMS values	1 week
Voltage Asymmetry	2 %	95 % of the 10-minute averages of RMS values of negative sequence voltages	1 week
Frequency (Grid-connected)	49.5 to 50.5 Hz	99.5 % of the 10-second average of the fundamental voltage component frequency	1 year
Frequency (Islanded grid)	47 to 52 Hz	100 % of the 10-second average of the fundamental voltage component frequency	1 year
Individual Harmonics	According to Figure A.1	95 % of the 10-minute averages values of each individual harmonic	1 week
Total Harmonics Distortion (THD)	8 %	95 % of the 10-minute averages values of THD	1 week
Flicker	1	95 % of the long-term flicker	1 week
Ripple Control Signals	Different limits for each frequency band	99 % of the 3-socond average of signal voltages	1 day

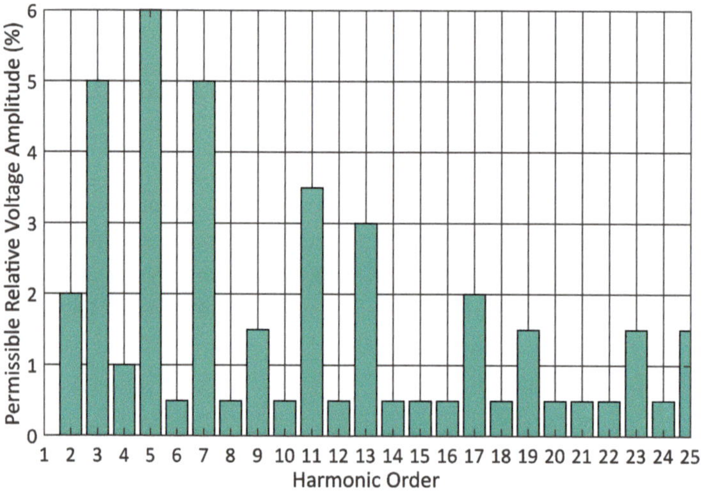

Figure A.1: Permissible relative voltage amplitudes of individual harmonics according the grid code DIN N 50160 [159].

Among the nine different low voltage distribution grids, at which measurements were conducted, no major power quality problem according to the criteria given in Table A.1 has been observed. As an example, the power quality indices measured at one of the low voltage distribution grids are presented in Table A.2.

Table A.2: Measured power quality indices at one of the low voltage distribution grids.

Electricity Supply Characteristic	**Acceptable Range/Values**	**Measured Values**		
		Phase	**95 % Value**	**Max.Value**
Voltage RMS	±10 %	L1	1.67 %	2.73 %
		L2	1.87 %	2.95 %
		L3	1.91 %	2.98 %
Voltage Asymmetry	2 %	All	0.17 %	0.21%
Total Harmonics Distortion (THD)	8 %	L1	1.99 %	2.21%
		L2	2.08 %	2.30 %
		L3	2.04 %	2.26 %
Flicker	1	L1	0.29	0.57
		L2	0.27	0.45
		L3	0.28	0.67
Frequency	49.5 to 50.5 Hz	**99.5% Value**	**100% Value**	**Max. Value**
		50.05 Hz	50.049 Hz	50.11 Hz

A.2 Summary of Interesting Recorded Incidents

In this section, some interesting observations from the recorded measurement data are presented, regarding voltage deviations, asymmetries, and voltage and current harmonics.

In terms of the voltage magnitude, a voltage deviation larger than ±10 % of the nominal value has been barely witnessed during the measurement period. A rare example of a voltage drop of more than 13 % is shown in Figure A.2. However, such voltage drops were far from being a power quality issue, as they were very rare and lasted for a very short period.

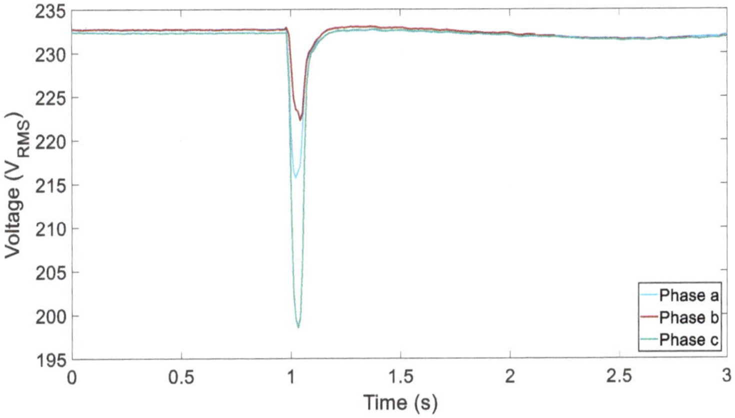

Figure A.2: An example of voltage drop above ±10% of the nominal value. Such voltage drops were extremely rare in the measurement and often for a noticeably short time.

However, voltage deviations above the ±3 % limit, from which the distributed energy resources and energy storage systems should support the grid according to the grid code VDE-AR-N 4105:2018-11 (see Figure 4.5(b)), has been recorded quite often. This has been especially the case in low voltage distribution grid with a high share PV generation system, where overvoltage incidents has been mostly observed. Therefore, it can be concluded that the reactive power support by the distributed energy resources is currently required often within each day at the measured locations. With the increasing share of PV in these areas, more voltage deviations in these areas are expected. An example of a recorded over-voltage incident above 3 % at one of the low voltage distribution grids is shown in Figure A.3.

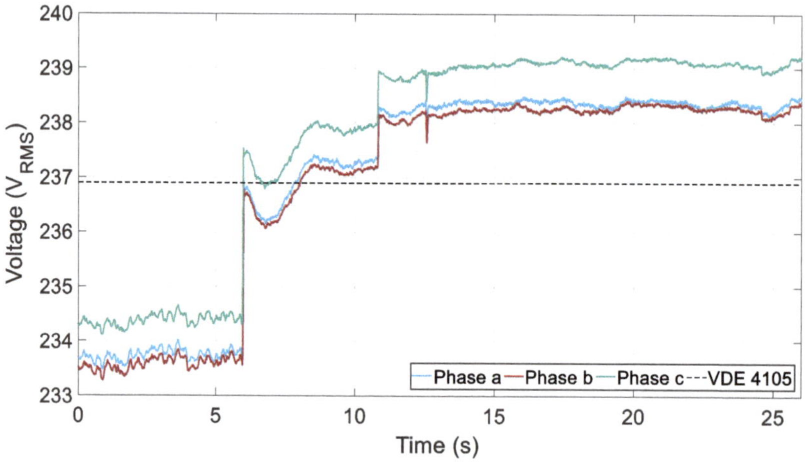

Figure A.3: An example of voltage deviations above ±3%, from which the distributed energy resources and energy storage system should support the grid with reactive power, according the German grid code VDE-ARN 4105:2018-11.

In terms of voltage and current asymmetry, voltage asymmetries were far below the power quality violation limit of 2 %, as discussed, reaching 0.67 % in the worst case at the measured locations. However, extremely high current asymmetries were recorded in almost all the measured low voltage grids, which has led to a high neutral current and negative sequence component. In some cases, the neutral current has been much greater than each individual phase currents, as illustrated in Figure A.4, as an example. Here, the negative sequence current has been multiple times more than the positive sequence current with the increase in the PV generation power around noon. This could be caused by old single-phase PV generation units, which increase the asymmetry in the grid around noon. It should be noted that grid connection of new single-phase PV system is no longer permitted in Germany since several years.

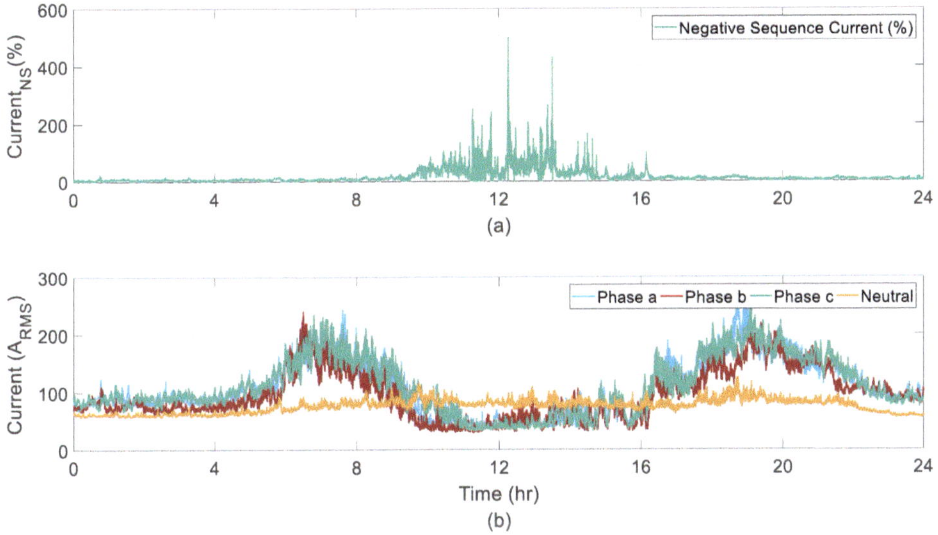

Figure A.4: An example of high current asymmetry during one day in one of the measured low voltage distribution grids. (a) Relative negative sequence current, (b) Phases and neutral currents.

Regarding the voltage and current harmonics, an interesting and yet expected observation is the increase in the THD of the measured voltages with the increase in the PV generation power around noon, at the low voltage grid with a high share of PV generation. This can be clearly seen in the example illustrated in Figure A.5. Here, the increase in demand during the day coincides with the increase the PV generation, leading to a reduced current drawn from the medium voltage grid and the local consumption of the PV generation around noon. Therefore, the current drawn from the medium voltage grid is reduced at this time. However, the voltage THD continuous to increase with the increase in the PV generation. While the this is still far away from the 8 % threshold to be a power quality concern, with the increase in the PV installations at these low voltage grids, harmonics can become an issue in the future. An example of currents drawn at such locations is shown in Figure A.6.

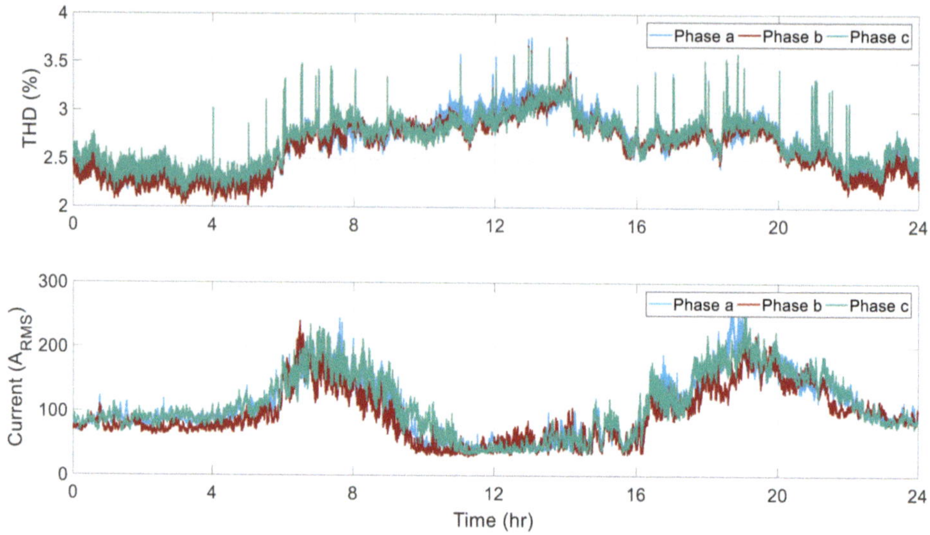

Figure A.5: An example of the increase in the THD of the voltage at noon with increase in the PV generation.

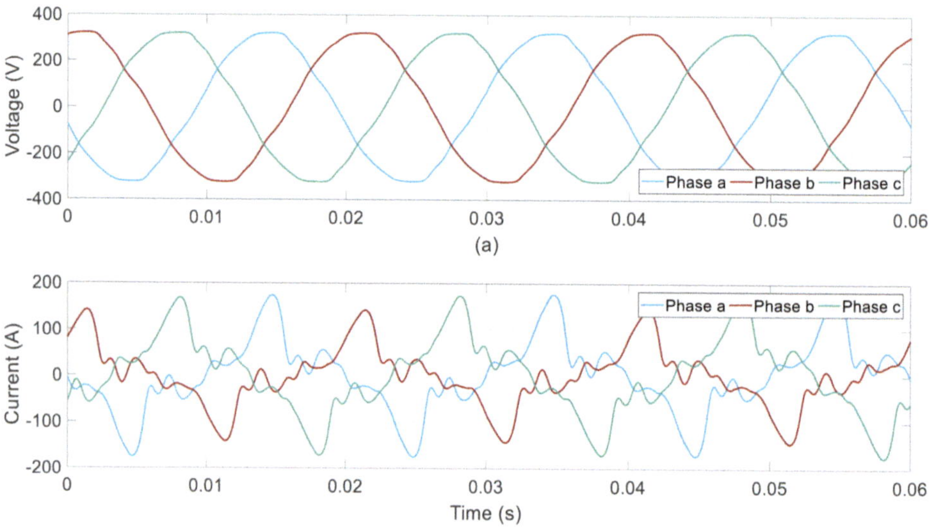

Figure A.6: An example of current and voltage waveforms at a low voltage distribution grid with a high THD.

Appendix B. Parameters of the Microgrid Components

The parameters of the distributed energy resources added to the CIGRE European low voltage distribution grid benchmark in Chapter 7 are presented in Tables B.1 and B.2.

Table B.1: Parameters of the diesel-based Combined Heat and Power (CHP) system.

Parameter	**Value**
Synchronous generator - 6th-order model	
Nominal apparent power	170 kVA
d-axis synchronous reactance (X_d)	1.305 pu
d-axis transient reactance (X'_d)	0.296 pu
d-axis subtransient reactance (X''_d)	0.252 pu
q-axis synchronous reactance (X_q)	0.474 pu
q-axis subtransient reactance (X''_q)	0.243 pu
Stator leakage reactance (X_l)	0.18 pu
d-axis transient open-circuit time constant (T'_{do})	4.49 s
d-axis subtransient open-circuit time constant (T''_{do})	0.0681 s
q-axis subtransient time constant (T''_q)	0.0513 s
Stator resistance (R_s)	0.003 pu
Inertia constant (H)	3.3 s
Governor	
Regulator gain (K)	29
Regulator time constant 1 (T_{r1})	0.01 s
Regulator time constant 2 (T_{r2})	0.02 s
Regulator time constant 3 (T_{r3})	0.2 s
Actuator time constant 1 (T_{a1})	0.25 s
Actuator time constant 2 (T_{a2})	0.009 s
Actuator time constant 3 (T_{a3})	0.0384 s
Engine time delay (T_{dl})	0.024 s
Automatic Voltage Regulator (AVR) – IEEE Type AC1A	
Voltage regulator Gain (K_a)	400
Voltage regulator time constant (T_a)	0.4 s
Damping filter gain (K_f)	0.03
Damping filter time constant (T_f)	1 s
Measurement low-pass filter time constant (T_r)	0.01 s
Transient gain reduction lag time constant (T_b)	0 s
Transient gain reduction lead time constant (T_c)	0 s

Table B.2: Parameters of the photovoltaic systems.

Parameter	Value
Photovoltaic system 1 (PV1)	
Nominal apparent power	50 kVA
DC-link voltage control - Proportional term	80
DC-link voltage control - Integral term	5000
d-axis current controller - Proportional term	5
d-axis current controller - Integral term	100
q-axis current controller - Proportional term	1
q-axis current controller - Integral term	10
DC-link capacitance	3 mF
PV voltage controller- Proportional term	1
PV voltage controller- Integral term	150
PV current controller- Proportional term	0.004
PV current controller- Integral term	1
Maximum power point tracking - Increment value	3 mV
Photovoltaic system 2 (PV2)	
Nominal apparent power	20 kVA
DC-link voltage control - Proportional term	80
DC-link voltage control - Integral term	1000
d-axis current controller - Proportional term	5
d-axis current controller - Integral term	100
q-axis current controller - Proportional term	5
q-axis current controller - Integral term	100
DC-link capacitance	3 mF
PV voltage controller- Proportional term	1
PV voltage controller- Integral term	150
PV current controller- Proportional term	0.004
PV current controller- Integral term	1
Maximum power point tracking - Increment value	3 mV

Table of Figures

Figure 2.1: Configuration of a conventional Flywheel Energy Storage System (FESS). 7

Figure 2.2: Comparison of the specific power and energy of the FESS with supercapacitors and lithium-ion (Li-ion) batteries (adopted from [29]). 8

Figure 2.3: Structure of the container holding the rotor and the electrical machine in a high-speed FESS (Picture courtesy: Stornetic GmbH [68], used with permission). 15

Figure 2.4: Flywheel connected to the common DC link of a wind generator for power smoothing applications. 18

Figure 2.5: Structure of a FESS with of High-temperature Superconductors (HTS) bearings. 20

Figure 3.1: The hard real-time constraint and the definition of an overrun. 24

Figure 3.2: Schematic diagram of a typical PHIL setup for distributed energy resources. 30

Figure 3.3: Real-time simulation and PHIL testing requirements according to the model bandwidth, defined as the maximum frequency of the harmonics or system dynamics to be represented accurately [126]. 32

Figure 3.4: (a) Field test setup (ideal case). (b) PHIL setup using the ideal transformer method. 33

Figure 3.5: Block diagram of a PHIL setup when using the ideal transformer method. 34

Figure 3.6: The influence of the system impedance ratio on the PHIL setup stability using the ideal transformer method. 35

Figure 3.7: The influence of the filter cut-off frequency on the PHIL setup stability using the ideal transformer method for a $ZHuTZeq$ ratio of 0.1 using the Nyquist criteria. 36

Figure 3.8: The influence of filter cut-off frequency on the power hardware-in-the-loop setup stability using the ideal transformer method for a $ZHuTZeq$ ratio of 0.1 using the Bode criteria. 37

Figure 3.9: The influence of simulation time step on the power hardware-in-the-loop setup stability using the ideal transformer method for a ZHuTZeq ratio of 0.1 using the Bode criteria. 38

Figure 3.10: Schematic diagram of the Damping Impedance Method (DIM). 38

Figure 3.11: Block diagram of PHIL setup when using the damping impedance method. 39

Figure 3.12: Overview of the 1 MVA PHIL setup at KIT's EnergyLab 2.0. 43

Figure 3.13: Inside of a 200 kVA COMPISO System Unit (CSU) power amplifier from Egston Power [126]. Cabinet 1 shows the grid connection and the isolation transformer, while the grid-side converter and its filter is in cabinet 2. The 6 COMPISO Digital Amplifier (CDA) are in cabinet 3, along with the Egston controller. Cabinet 4 shows the output contactors for the hardware under test. 44

Figure 3.14: Five 200 kVA GAMP6 Egston switched-mode power amplifiers (1 MVA in total). 44

Figure 4.1: The single-line diagram of a simplified power system for calculating the voltage sensitivity. 49

Figure 4.2: Different voltage sensitivity coefficients in (a) high voltage transmission grids, and (b) low voltage distribution grids (figures adopted from [170]). 51

Figure 4.3: Examples of notations used for estimating the mutual information using the KSG estimate. 55

Figure 4.4: The flow chart of the proposed allocation algorithm for the FESS based on mutual information. 57

Figure 4.5: Active and reactive power control characteristics according to the German grid code VDE-AR-N 4105:2018-11. (a) The P(f) characteristic for frequency deviations, and (b) the Q(U) characteristic for voltage deviations. 58

Figure 4.6: Schematic example of a measurement setup at the low voltage distribution networks. 59

Figure 4.7: Comparison of the mutual information (MI) scores of the top nine candidates for the FESS installation. 61

Figure 4.8: Measurement data at bus 4 and 5 illustrated for a period of two days with a 1-minute data resolution. (a) Bus 4 voltage, (b) bus 4 active power, (c) bus 5 voltage, and (d) bus 5 active power. Bus 4 has the highest mutual information score between voltage and active power, while bus 5 has the lowest score. 62

Figure 4.9: An example of the variations in the voltages and currents at bus 1, the most attractive candidate for the FESS installation during a period of 6 hours. 63

Figure 4.10: Example of a daily power profile of a low voltage distribution grid with a high share of PV generation during a summer day of 2018 in southern Germany 64

Figure 4.11: Active power measurements at (a) bus 1, (b) bus 2, (c) bus 3, and (d) bus 4. The data is shown with the resolution of 1 minute for a period of 8 days, starting from Monday, which makes the 6th and 7th days, weekends. 69

Figure 4.12: One-sided frequency spectrum of the power profiles at all buses using the 1-second data for a period of one week. 70

Figure 4.13: The process of the motif discovery using SAX and random projection (adopted from [207]). 71

Figure 4.14: An example of PAA of a time series (black line) and its SAX representation (the blue letters). The time series itself is shown with the grey line (adopted from [207]). 72

Figure 4.15: The standard consumption patterns at the four buses: (a) Bus 1: No PV generation with industrial loads. (b) Bus 2: Limited PV generation to cover only the increase in consumption at peak generation. (c) Bus 3: Significant share of PV generation in a residential grid, resulting in a typical *duck curve*. (d) Bus 4: High share of PV production in a residential grid, leading to slight negative power flow at peak generation. 74

Figure 4.16: The spectrogram of the standard patterns using the 1-second data and hamming window of 1-minute length with no overlaps for (a) bus 1, (b) bus 2, (3) bus 3, and (d) bus 4. 75

Figure 4.17: The three different patterns detected at bus 2 using motif discovery. Pattern 1 is the consumption pattern during a typical workday with the highest number of reoccurrences, while pattern 2 shows the consumption pattern over the weekdays and public holidays. Pattern 3 represents the power profile of the first day after a weekend or a holiday. 76

Figure 4.18: Changes in the objective function with the increase of the cut-off frequency of the filter at each bus using the data with 1-second resolution. 81

Figure 4.19: The allocated power to each type of ESS at bus 1 by applying a low-pass filter with the optimum cut-off frequency. 81

Figure 4.20: (a) The maximum ramp rate, and (b) the number of mode changes per day for both types of ESS using the 1-second data. 83

Figure 4.21: The variations in the characteristics of the type 1 and type 2 ESS with respect to the change in the cut-off frequency of the low-pass filter for bus 2. The optimum cut-off frequency is shown with the red marker. 84

Figure 4.22: The active power drawn from the medium voltage grid at bus 3, with and without the type 1 ESS, and in combination with a type 2 ESS, such as a FESS, (a) during the whole two weeks, (b) during the 5th day, and (c) between 22:00 and 23:00 of the 5th day. 87

Figure 4.23: The state of charge variations during the two weeks for both ESS types, with the characteristics derived from the standard patterns. 87

Figure 5.1: Structure of a high-speed Flywheel Energy Storage System (FESS). 93

Figure 5.2: The inside view of the container of the 60 kW high-speed FESS. 93

Figure 5.3: Arrangement of the permanent magnets in (a) surface mounted PMSM, (b) surface inset PMSM, (c) interior (I-shaped) PMSM, and (d) interior circumferential PMSM. 95

Figure 5.4: (a) The equivalent circuit of a PMSM in the three-phase stationary reference frame. (b) Schematic of a two-pole surface mounted PMSM in the dq-rotating reference frame. 95

Figure 5.5: Equivalent circuits of a PMSM in the (a) d-axis and the (b) q-axis. 98

Figure 5.6: The equivalent circuit of the average model of a voltage source converter. 102

Figure 5.7: The equivalent circuit of the switching model of a voltage source converter. 103

Figure 5.8: (a) Block diagram of the Grid-side Converter (GSC) controller. The upper part of the controller is responsible for supporting the grid voltage, while the lower part controls the DC-link voltage. (b) Block diagram of the Machine-Side Converter (MSC) controller. The upper part is responsible for the providing frequency support to the grid. The Maximum Torque per Ampere (MTPA) control is implemented for controlling the PMSM....................108

Figure 5.9: Different control strategies for controlling a PMSM (figure adopted from [247]).....................110

Figure 5.10: Vector diagram of the current and the voltage of the PMSM (a) without the maximum torque per ampere control, and (b) with the maximum torque per ampere control.....................111

Figure 5.11: Block diagram of a synchronous reference frame phase-locked loop with a low-pass filter.....................113

Figure 5.12: The single-line diagram of the simulated system.....................115

Figure 5.13: The active power injected by the FESS during different steps in the grid frequency.....................116

Figure 5.14: Real-time and non-real-time simulation results during a step in the grid frequency by 1.2 Hz.....................117

Figure 5.15: Real-time and non-real-time simulation results during a 7% voltage sag.....................118

Figure 6.1: The schematic diagram of the setup used for the PHIL testing of the 60 kW high-speed FESS.....................120

Figure 6.2: The measured and simulated active power of the FESS during the PHIL testing of scenario A, charging the FESS at the rated power.....................123

Figure 6.3: The measured and simulated state of charge of the FESS during the PHIL testing of scenario A, charging the FESS at the rated power.....................124

Figure 6.4: The variations in the grid frequency for grid code compliance validation of the high-speed FESS, adopted from the grid code VDE V 0124-100:2020-06 [257].....................125

Figure 6.5: The measured and simulated grid frequency during the PHIL testing of scenario B, grid code compliance verification test.....................126

Figure 6.6: The measured and simulated active power of the FESS during the PHIL testing of scenario B, grid code compliance verification test. ... 127

Figure 6.7: The measured and simulated state of charge of the FESS during the PHIL testing of scenario B, grid code compliance verification test. ... 127

Figure 6.8: The response time of the high-speed FESS to a step in the grid frequency during the PHIL testing. ... 128

Figure 6.9: Sequence of event involved in UK's major frequency incident of 9 August, 2019 [258]. ... 129

Figure 6.10: The measured and simulated grid frequency during the PHIL testing of scenario C, response to the frequency incident of August 9, 2019 in the UK. ... 130

Figure 6.11: The measured and simulated active power of the FESS during the PHIL testing of scenario C, response to the frequency incident of August 9, 2019 in the UK. ... 131

Figure 6.12: The measured and simulated state of charge of the FESS during the PHIL testing of scenario C, response to the frequency incident of August 9, 2019 in the UK. ... 131

Figure 6.13: The single-line diagram of the real low voltage distribution network in southern Germany, used for PHIL testing of the FESS in scenario D. ... 132

Figure 6.14: The PHIL testing results of scenario D, increasing the frequency support of low-voltage grids. (a) Power of the PV systems, (b) the measured and simulated grid frequency, (c) The total power drawn by the low voltage (LV) grid, and (d) the measured and simulated power of the FESS. ... 135

Figure 6.15: The measured and simulated state of charge of the FESS during the PHIL testing of scenario D, increasing the frequency support of low-voltage grids. ... 136

Figure 7.1: Typical behavior of the frequency in a power system after a major large loss in generation (adopted from [266]). ... 138

Figure 7.2: Simplified block diagram of frequency control loops in a simplified power system with the addition of emulated inertia (model according to [267]). ... 139

Figure 7.3: The effect of increasing the emulated inertia constant using converter-interfaced systems on (a) the frequency, and (b) the ROCOF during an under-frequency event. 140

Figure 7.4: The pole-zero map of the simplified power system with the increasing value of the emulated inertia constant (H). 140

Figure 7.5: The block diagram of the proposed inertia emulation controller, and the State of Charge (SOC) controller. This controller is to be added the MSC controller of the FESS in Chapter 5. 147

Figure 7.6: Segmentation of the frequency deviation period into the accelerating phase (red) and the decelerating phase (yellow), according to the sign of $\Delta\omega gd\Delta\omega gdt$. 149

Figure 7.7: The variations of the inertia emulation control parameters with respect to the speed of the FESS. (a) The inertia constants, $H1$ and $H2$, (b) the damping coefficients $D1$ and $D2$, and (c) the adaptive coefficients KH and KD. 153

Figure 7.8: The block diagram of the Dual Second-Order Generalized Integrator Frequency-Locked Loop (DSOGI-FLL) (adopted from [304]). 154

Figure 7.9: The single-line diagram of the test microgrid based on the CIGRE European low voltage network benchmark. 156

Figure 7.10: Simulation results for scenario A: microgrid islanding. (a) Frequency, (b) ROCOF, (c) Inertia constant (H), and (d) Damping coefficient (D). 159

Figure 7.11: Simulation results for scenario A: microgrid islanding. (a) FESS Power, and (b) FESS State of Charge (SOC). 160

Figure 7.12: Simulation results for scenario B: drop in PV generation. (a) Frequency, (b) ROCOF, (c) Inertia constant (H), and (d) Damping coefficient (D). 162

Figure 7.13: Simulation results for scenario B: drop in PV generation. (a) FESS Power, and (b) FESS State of Charge (SOC). 162

Figure 7.14: The signal connections and PHIL setup for implementing the inertia emulation controller on the FESS using the concept of rapid control prototyping. 164

Figure 7.15: Response of the FESS to steps in the active power reference, sent from the OP5700 real-time simulator. (a) A 60 kW step, from steady-state operation and no power injection. (b) A 120 kW step, from charging at the nominal power to discharging at the same rate. 165

Figure 7.16: PHIL testing results for scenario A: microgrid islanding. (a) Frequency, (b) ROCOF, (c) Inertia constant (H), and (d) Damping coefficient (D). 167

Figure 7.17: PHIL testing results for scenario A: microgrid islanding. (a) Measured FESS power, (b) FESS State of Charge (SOC). 168

Figure 7.18: PHIL testing results for scenario B: drop in PV generation. (a) Frequency, (b) ROCOF, (c) Inertia constant (H), and (d) Damping coefficient (D). 170

Figure 7.19: PHIL testing results for scenario B: drop in PV generation. (a) Measured FESS power, (b) FESS State of Charge (SOC). 170

Figure 7.20: A comparison of the real FESS behavior and its simulated model, given the same active power reference from the PHIL testing 171

Figure A.1: Permissible relative voltage amplitudes of individual harmonics according the grid code DIN N 50160 [159]. 208

Figure A.2: An example of voltage drop above ±10% of the nominal value. Such voltage drops were extremely rare in the measurement and often for a noticeably short time. 209

Figure A.3: An example of voltage deviations above ±3%, from which the distributed energy resources and energy storage system should support the grid with reactive power, according the German grid code VDE-ARN 4105:2018-11. 210

Figure A.4: An example of high current asymmetry during one day in one of the measured low voltage distribution grids. (a) Relative negative sequence current, (b) Phases and neutral currents. 211

Figure A.5: An example of the increase in the THD of the voltage at noon with increase in the PV generation. 212

Figure A.6: An example of current and voltage waveforms at a low voltage distribution grid with a high THD. 212

Table of Tables

Table 2.1: Advantages and disadvantages of a FESS in comparison to other storage technologies.......8

Table 2.2: A comparison of the characteristics of several commercial FESS [34].......9

Table 2.3: Applications of FESS in different sectors and fields. 10

Table 2.4: Comparison of the main two types of FESS [66], [67].......14

Table 2.5: Characteristics of several materials used for the rotor in a FESS [28], [66].......16

Table 2.6: Comparison of different types of electrical machines used in FESS [53], [78].......17

Table 2.7: Comparison of different bearing technologies used in FESS.......20

Table 3.1: Possible techniques and solutions for real-time simulations of low voltage distribution grids with a high share of distributed energy resources. 28

Table 3.2: Different testing concepts using real-time simulations.......29

Table 3.3: A step-by-step guideline for PHIL testing for grid integration studies [118].......31

Table 3.4: The parameters used for the PHIL stability analysis.......36

Table 3.5: A comparison of the most commonly used interfacing algorithms for PHIL testing [98], [129], [138]. 40

Table 3.6: Literature review on PHIL testing of energy storage systems. 42

Table 3.7: The characteristics of each Egston amplifier at KIT-ITEP.......44

Table 4.1: Summary of the previously used methods for reducing data size for sizing studies and their drawbacks and limitations. 66

Table 4.2: The optimum cut-off frequency of the low-pass filter for each bus using different data resolutions. 81

Table 4.3: Sizing results according to the standard patterns at each bus for the three different data resolutions. 82

Table 4.4: Sizing results using three different standard at bus 2 using the 1-minute data. 85

Table 5.1: A comparison between the surface mounted and interior PMSM [231]. 94

Table 5.2: Parameters of the PMSM of the high-speed FESS according to the manufacturer. 100

Table 5.3: Variables definition and parameters of the grid-side converter (GSC) controller. 107

Table 5.4: Variables definition and parameters of the machine-side converter (MSC) controller. 111

Table 5.5: Variables definition and parameters of the synchronous reference frame phase-locked loop (SRF-PLL). 113

Table 5.6: Parameters of the total system losses and auxiliary power requirements. 115

Table 6.1: List of the most important commands and measurement values communicated through the Ethernet connection using the Modbus TCP/IP protocol between the FESS controller and the real-time simulator. 121

Table 6.2: Comparison of the model execution times of the German low voltage distribution grid using different techniques used for reducing the model execution time. 134

Table 7.1: The parameters used for the simplified single-machine power system. 139

Table 7.2: Technical limitations of different active power sources for inertia emulation. 143

Table 7.3: A comparison of the proposed inertia emulation control with several interval-based methods in literature. 150

Table 7.4: The parameters used for the proposed inertia emulation controller. 153

Table 7.5: A summary of the controller characteristics in all cases. 157

Table 7.6: The maximum ROCOF and frequency nadir for all the simulated scenarios and cases. 160

Table A.1: Summary of the power quality requirements at the low voltage level in Germany [159], [307]. 208

Table A.2: Measured power quality indices at one of the low voltage distribution grids. 209

Table B.1: Parameters of the diesel-based Combined Heat and Power (CHP) system. 213

Table B.2: Parameters of the photovoltaic systems. 214

Abbreviations

BESS	Battery Energy Storage System
CHP	Combined Head and Power
DER	Distributed Energy Resources
DSOGI	Double Second-Order Generalized Integrator
ESS	Energy Storage System
FESS	Flywheel Energy Storage System
FLL	Frequency-locked Loop
GSC	Grid-side Converter
ITM	Ideal Transformer Method
LV	Low Voltage
MI	Mutual Information
MSC	Machine-side Converter
PHIL	Power Hardware-in-the-Loop
PMSM	Permanent Magnet Synchronous Machine
ROCOF	Rate of Change of Frequency
SOC	State of Charge
SRF-PLL	Synchronous Reference Frame Phase-Locked Loop